WJEC

GCSE
Applied Science

Single and Double Award

Adrian Schmit, Jeremy Pollard, Sam Holyman

HODDER
EDUCATION
AN HACHETTE UK COMPANY

Although every effort has been made to ensure that website addresses are correct at time of going to press, Hodder Education cannot be held responsible for the content of any website mentioned in this book. It is sometimes possible to find a relocated web page by typing in the address of the home page for a website in the URL window of your browser.

Hachette UK's policy is to use papers that are natural, renewable and recyclable products and made from wood grown in well-managed forests and other controlled sources. The logging and manufacturing processes are expected to conform to the environmental regulations of the country of origin.

Orders: please contact Hachette UK Distribution, Hely Hutchinson Centre, Milton Road, Didcot, Oxfordshire, OX11 7HH. Telephone: +44 (0)1235 827827. Email education@hachette.co.uk Lines are open from 9 a.m. to 5 p.m., Monday to Friday. You can also order through our website: www.hoddereducation.co.uk

ISBN: 978 1 3983 6903 0

© Sam Holyman, Adrian Schmit and Jeremy Pollard 2022

First published in 2022 by
Hodder Education,
An Hachette UK Company
Carmelite House
50 Victoria Embankment
London EC4Y 0DZ

www.hoddereducation.co.uk

Impression number   10 9 8 7 6 5 4 3 2 1
Year   2026 2025 2024 2023 2022

Cover photo © Meaw_stocker-stock.adobe.com
Typeset in India by Integra Software Services Pvt. Ltd.
Printed in Italy.

A catalogue record for this title is available from the British Library.

Contents

Get the most from this book

Welcome to the WJEC GCSE Applied Science Student Book.

This book covers all of the Foundation and Higher-tier content for the 2016 WJEC GCSE Applied Science Single and Double Award specifications.

The following features have been included to help you get the most from this book.

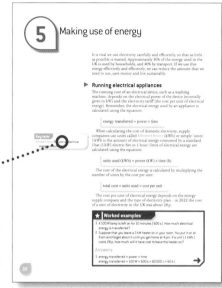

Most of the content in this book is suitable for all students. However, some chapters should only be studied by those taking WJEC GCSE Applied Science: Double Award. This content is clearly marked by a green line on the pages. Students taking WJEC GCSE Applied Science: Single Award do not need to study these pages.

Key terms

Important words and concepts are highlighted in the text and clearly explained for you in the margin.

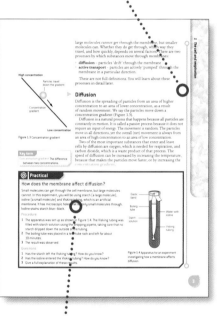

Activity

These activities usually involve the use of second-hand data that could not be obtained in the school laboratory, along with questions that will test your scientific enquiry skills.

Practical

Full practical guidance is not included in these boxes; you will complete specified practicals in class under the direction of your teacher. These activities will help consolidate your learning and test your understanding of practical skills. Completing these will help you prepare for questions on practical work that come up in the exam

Discussion point

These are questions that could be answered by individuals, but that benefit from discussion with your teacher or others in your class. In such cases there are usually a variety of opinions or possible answers to explore.

Test yourself

These short questions, found throughout each chapter, allow you to check your understanding as you progress through a topic.

Practice questions

You will find practice questions at the end of every chapter. These follow the style of the different types of questions you might see in your examination and have marks allocated to each question part.

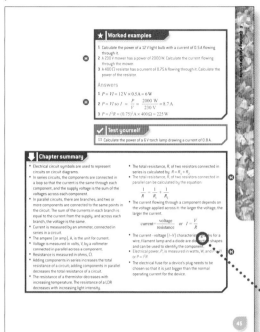

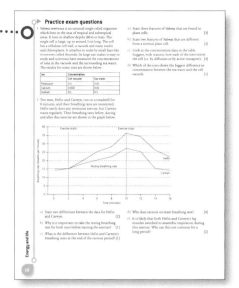

Chapter summary

This provides an overview of everything you have covered in a chapter and is a useful tool for checking your progress and for revision.

Some material in this book is only required for students taking the Higher-tier examination. This content is clearly marked by the Higher icon.

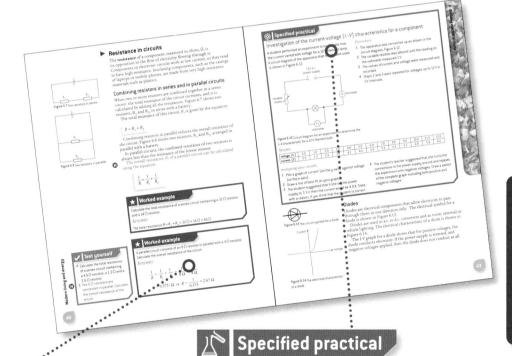

Answers

Answers for all questions and activities in this book can be found online at: www.hoddereducation.co.uk/wjecappliedscience

Specified practical

WJEC's specified practicals are clearly highlighted. Full practical guidance is not included in these boxes; you will complete specified practicals in class under the direction of your teacher. These activities will help consolidate your learning and test your understanding of practical skills. Completing these will help you prepare for questions on practical work that come up in the exam.

Worked example

Examples of questions and calculations that feature full workings and sample answers.

Acknowledgements

Page 7 © Biophoto Associates/Science Photo Library; page 26 © Paul Glendell / Alamy Stock Photo; page 27 © bilanol/stock.adobe.com; page 36 © Haydn Denman / Alamy Stock Photo; page 57 *t* © MARTYN F. CHILLMAID / SCIENCE PHOTO LIBRARY, b © ANDREW LAMBERT PHOTOGRAPHY / SCIENCE PHOTO LIBRARY; page 70 © Martin Shields / Alamy Stock Photo; page 71 © MARTYN F. CHILLMAID/SCIENCE PHOTO LIBRARY; page 80 *l t-b* © NASA/Johns Hopkins University Applied Physics Laboratory/Carnegie, © NASA/JPL-Caltech, © NASA, © NASA, ESA, and STScI, © NASA, ESA, A. Simon (Goddard Space Flight Center), and M.H. Wong (University of California, Berkeley), © NASA, ESA, A. Simon (GSFC), M.H. Wong (University of California, Berkeley) and the OPAL Team, © NASA/Space Telescope Science Institute, © ESO/P. Weilbacher (AIP), *r t-b* © NASA, © NASA/SDO; page 83 © NASA/SDO; page 87 © lucky-photo/stock.adobe.com; page 90 © Bob/stock.adobe.com; page 91 © pololia/stock.adobe.com; page 108 *l-r* © Premaphotos / Alamy Stock Photo, © Christopher Taylor / Alamy Stock Photo, © agefotostock / Alamy Stock Photo, 06_06d_1597px-Lecanora_conizaeoides_a1_(8)_Wiki; page 118 PAUL RAPSON / SCIENCE PHOTO LIBRARY; page 122 t © GE MEDICAL SYSTEMS / SCIENCE PHOTO LIBRARY, b © Thomas Heitz/stock.adobe.com; page 123 *l t-b* © Ludmila/stock.adobe.com, © DOUGLAS/stock.adobe.com, © STEVIE GRAND/SCIENCE PHOTO LIBRARY, *r* © PROF. J. LEVEILLE/SCIENCE PHOTO LIBRARY; page 127 © Mark Kostich/stock.adobe.com; page 129 © ALVIN TELSER / SCIENCE PHOTO LIBRARY; page 132 © SCIENCE PHOTO LIBRARY; page 135 © SutthaB/stock.adobe.com; page 149 © Cultura Creative Ltd / Alamy Stock Photo; page 153 *t* © RHJ/stock.adobe.com, *b* © ijp2726/stock.adobe.com; page 163 © nottsutthipong/stock.adobe.com; page 172 © BillionPhotos.com/stock.adobe.com; page 176 © MARTYN F. CHILLMAID / SCIENCE PHOTO LIBRARY; page 180 © FUNDAMENTAL PHOTOS / SCIENCE PHOTO LIBRARY; page 185 *t* © ANDREW LAMBERT PHOTOGRAPHY / SCIENCE PHOTO LIBRARY, *b* © OONA STERN / SCIENCE PHOTO LIBRARY

t = top, *b* = bottom, *l* = left, *m* = middle, *r* = right

Data to create the image on page 206 licenced under CC-BY from OurWorldinData.org

Data sources: Markandya & Wilkinson (2007); UNSCEAR (2008; 2018); Sovacool et al. (2016); IPCC AR5 (2014); Pehl et al. (2017) and Ember Energy (2021).

Every effort has been made to trace all copyright holders, but if any have been inadvertently overlooked, the Publisher will be pleased to make the necessary arrangements at the first opportunity.

1 The cell and respiration

Cells are the basic 'unit' of all living things. Cells were first seen through a microscope and described by Robert Hooke (1635–1703) in 1665.

Although all cells have features in common, there are also differences. Some of those differences allow scientists to classify cells as either animal cells or plant cells.

▶ Plant and animal cells

All cells in both plants and animals have certain features in common:

- **cytoplasm** – a 'living jelly' where most of the cell's chemical reactions go on
- **cell membrane** – surrounds the cytoplasm and controls what enters and leaves the cell
- **nucleus** – contains **DNA**, the chemical which controls the cell's activities.

Plant cells can be distinguished from animal cells because they have some extra features:

- a **cell wall** – made of cellulose, this surrounds all plant cells
- **central vacuole** – a large, permanent space filled with liquid cell sap
- **chloroplasts** – absorb the light plants need to make their food by photosynthesis and are found in all plant cells but never in animal cells.

Figure 1.1 shows examples of plant and animal cells, and the differences between them.

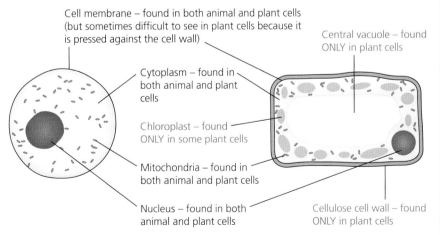

Figure 1.1 Animal cell (left) and plant cell (right) showing differences in structure

sperm cell

red blood cell

xylem cell

Figure 1.2 Specialised cells

Table 1.1 summarises the information you need to know about the structure of cells.

Table 1.1 A summary of cell structure

Organelle	Where found	Function
Nucleus	All cells	Contains DNA, which controls the cell's activities
Cell membrane	All cells	Controls what enters and leaves the cell
Cytoplasm	All cells	Forms the bulk of the cell and is where most of the chemical reactions occur
Chloroplasts	Some plant cells	Absorb light for photosynthesis
Cell wall	Plant cells	Supports the cell
Vacuole	Plant cells	Filled with a solution of nutrients including glucose, amino acids, and salts

Cell specialisation

The first living organisms were single cells that carried out all life functions. Over time, organisms became **multicellular**, and cells were specialised for particular functions (Figure 1.2). This means some cells look very different to the examples in Figure 1.1.

Levels of organisation

During the development of an animal or plant, the cells are organised into groups called **tissues**. Different tissues are grouped together to form **organs**, and organs may have linked functions in **organ systems**. Definitions and examples of the different levels of organisation are shown in Table 1.2. An organism is the scientific term for *any* living thing – it may not actually have organs.

Table 1.2 Levels of organisation in the structure of living things

Level of organisation	Definition	Examples
Tissue	A group of similar cells with similar functions	Bone, muscle, blood, xylem, epidermis
Organ	A collection of two or more tissues with specific functions	Kidney, brain, heart, leaf, flower
Organ system	A collection of several organs that work together	Digestive, nervous, respiratory, shoot and root systems

✓ Test yourself

1 State three features that are found in both animal and plant cells.
2 State three features that are found in plant cells but not in animal cells.
3 What is the function of the cell wall in plant cells?
4 Suggest a reason many plant cells do not contain chloroplasts.
5 At what level of organisation (cell, tissue, or organ) is the heart?

Movement into and out of cells

To get into and out of cells, substances must pass through the cell membrane. The cell membrane is **selectively permeable**, which means it lets some molecules through but not others. In general, large molecules cannot get through the membrane, but smaller molecules can. Whether they do get through, which way they travel, and how quickly, depends on several factors. There are two processes by which substances move through membranes:

▸ **diffusion** – particles 'drift' through the membrane
▸ **active transport** – particles are actively 'pumped' through the membrane in a particular direction.

These are not full definitions. You will learn about these processes in detail later.

▷ Diffusion

Diffusion is the spreading of particles from an area of higher concentration to an area of lower concentration, as a result of random movement. We say the particles move down a concentration gradient (Figure 1.3).

Diffusion is a natural process that happens because all particles are constantly in motion. It is called a passive process because it does not require an input of energy. The movement is random. The particles move in all directions, yet the overall (net) movement is always from an area of high concentration to an area of low concentration.

Two of the most important substances that enter and leave cells by diffusion are oxygen, which is needed for respiration, and carbon dioxide, which is a waste product of that process. The speed of diffusion can be increased by increasing the temperature, because that makes the particles move faster, or by increasing the concentration gradient.

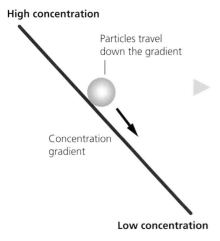

High concentration

Particles travel down the gradient

Concentration gradient

Low concentration

Figure 1.3 Concentration gradient

Key term

Concentration gradient The difference between two concentrations.

⚙ Practical

How does the membrane affect diffusion?

Small molecules can get through the cell membrane, but large molecules cannot. In this experiment, you will be using starch (a large molecule), iodine (a small molecule) and Visking tubing, which is an artificial membrane. It has microscopic holes that let only small molecules through. Iodine stains starch blue–black.

Procedure

1 The apparatus was set up as shown in Figure 1.4. The Visking tubing was filled with starch solution using the dropping pipette, taking care that no starch dripped down the outside of the tubing.
2 The boiling tube was placed in a test-tube rack and left for about 10 minutes.
3 The result was observed.

Questions

1 Has the starch left the Visking tubing? How do you know?
2 Has the iodine entered the Visking tubing? How do you know?
3 Give a full explanation of these results.

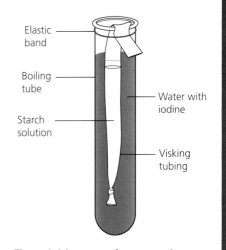

Elastic band

Boiling tube

Starch solution

Water with iodine

Visking tubing

Figure 1.4 Apparatus for an experiment investigating how a membrane affects diffusion

Active transport

Diffusion transports substances down a concentration gradient. Sometimes, cells need to get particles into or out of the cytoplasm against a concentration gradient, from an area of lower concentration to an area of higher concentration. This will not happen by diffusion. To move the particles, the cell must use energy to 'pump' the particles against the concentration gradient. As this type of transport requires an input of energy, it is called active transport.

✔ Test yourself

6 Diffusion is a *passive* process. What does the term *passive* mean?
7 What is a concentration gradient?
8 Why do sugar molecules get through the cell membrane, but starch molecules do not?
9 Why is it important to use the term *net* movement when describing diffusion?
10 Why is active transport needed in cells?

Aerobic respiration

All cells need energy. This is produced by the breakdown of food molecules, which store chemical energy. **Respiration** is the process in which the food is broken down and the energy is released for use. If a cell stops respiring, it dies. Many other processes are needed for life, but respiration is continuous for the whole of the organism's life.

The food molecule that is usually respired is glucose (although it is possible to use others). Most respiration is aerobic and oxygen is used up in the process. Carbon dioxide and water are produced as waste materials.

The word equation for aerobic respiration is:

glucose + oxygen → carbon dioxide + water + ENERGY

The equation is an over-simplification. Aerobic respiration is a *series* of chemical reactions, controlled by enzymes. A number of different factors, including temperature and pH, affect the rate of respiration in a cell.

Key term

Enzyme Biological molecule which acts as a catalyst, speeding up a chemical reaction but not taking part in it.

Anaerobic respiration

Cells do not always have a supply of oxygen. Some organisms live in places that are anaerobic (without oxygen) or where oxygen levels are extremely low. In humans and other mammals, oxygen levels in certain tissues can get very low (for example, in muscle tissue during strenuous exercise), yet these cells survive.

They survive because they can respire anaerobically. Even without oxygen, certain cells can partially break down glucose and release some of the energy from it.

In anaerobic respiration in animals, glucose is broken down into lactic acid (sometimes called lactate) and the word equation is simple:

glucose → lactic acid + ENERGY

Anaerobic respiration is much less efficient than aerobic respiration because the glucose is not fully broken down and much less energy is released for each molecule of glucose used. For this reason, animal cells always respire aerobically when they can, and only use anaerobic respiration when oxygen is in short supply.

Oxygen debt

If you run fast, you get breathless. When you stop, you breathe faster and deeper while you pay back your '**oxygen debt**'. During vigorous exercise, your breathing cannot supply your muscles with all the oxygen they need, so they switch to anaerobic respiration. As a result, lactic acid builds up. It can make your muscles ache, and there is still a lot of energy locked up in it (remember that glucose is not fully broken down in anaerobic respiration). Oxygen breaks down lactic acid and releases the remaining energy. So, when you finish the exercise, your body breathes faster and deeper to provide extra oxygen to break down the lactic acid. In effect, you are breathing in the oxygen that you needed (but could not get) during the exercise. You built up an oxygen debt, which is then repaid.

✔ Test yourself

11 Why will a cell die if it stops respiring?
12 What type of respiration uses oxygen?
13 Why can't human muscles use anaerobic respiration for a long period of time?
14 Why do people breathe deeper and faster after intense exercise?

Investigation of the factors that affect the rate of respiration

Students investigated the effect of glucose on the rate of respiration in yeast, a microscopic fungus. During respiration, bubbles of carbon dioxide are produced.

Procedure

1 $10\,cm^3$ of yeast suspension were measured into the boiling tube.
2 $10\,cm^3$ of 2% glucose solution were added.
3 Stirred with a stirring rod.
4 A few drops of oil on the top of the liquid were added using a pipette. It should have formed a layer over the surface.
5 The apparatus was assembled as shown in Figure 1.5.
6 The stopwatch was started when the first bubble appeared, and then the total number of bubbles produced in 2 minutes were counted.
7 Steps 1–6 were repeated with 4, 6, 8 and 10% glucose solution.
8 A line graph of concentration of glucose against number of bubbles per 2 minutes was plotted.

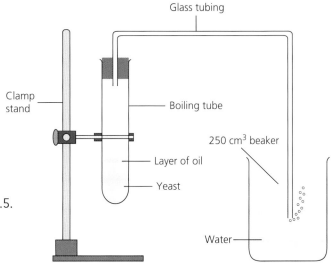

Figure 1.5 Experimental setup

Analysing the results

1 What are your conclusions about the effect of temperature on the rate of respiration?
2 Explain why this method is a fair test.
3 Explain how you would vary this method to test for the effect of temperature on respiration, instead of glucose concentration.

- Animal and plant cells have the following parts: cell membrane, cytoplasm, nucleus; in addition, plants cells have a cell wall, vacuole and sometimes chloroplasts.
- Cells differentiate in multicellular organisms to become specialised cells, adapted for specific functions.
- Tissues are groups of similar cells with a similar function; organs may comprise several tissues performing specific functions.
- Diffusion is the passive movement of substances down a concentration gradient.
- The cell membrane forms a selectively permeable barrier, allowing only certain substances to pass through by diffusion, most importantly oxygen and carbon dioxide.
- Visking tubing can be used as a model of a cell membrane.

- Active transport is an active process by which substances can enter cells against a concentration gradient.
- Aerobic respiration is a series of enzyme-controlled reactions that occur in cells when oxygen is available.
- Aerobic respiration uses glucose and oxygen to release energy, and produces carbon dioxide and water.
- Anaerobic respiration occurs when oxygen is not available. In animals, glucose is broken down into lactic acid/lactate.
- During strenuous exercise, anaerobic respiration in muscles builds up an oxygen debt, which is repaid after the exercise by breathing faster and deeper than normal.
- Anaerobic respiration is less efficient than aerobic respiration.

2 Obtaining the materials for respiration

▶ The respiratory system

The function of the respiratory system is to extract oxygen from the air and move it into the blood, so it can travel to all cells in the body. The respiratory system also removes carbon dioxide, which is a waste product of respiration.

▶ Structure of the respiratory system

The respiratory system of a human is shown in Figure 2.1.

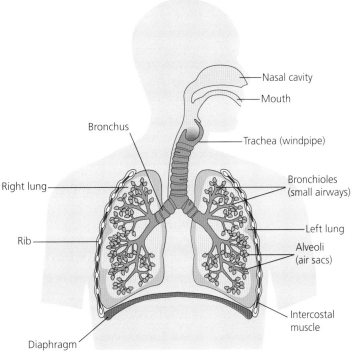

Nasal cavity

Mouth

Bronchus

Trachea (windpipe)

Bronchioles (small airways)

Right lung

Left lung

Alveoli (air sacs)

Rib

Intercostal muscle

Diaphragm

Figure 2.1 The human respiratory system

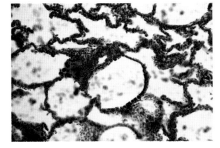

Figure 2.2 Microscopic section of lung tissue. The lungs are sponge-like and mostly composed of air

Air travels into the body via the nose and mouth. It enters the lungs through the **trachea**, which splits into two **bronchi** (singular: bronchus), one going to each lung. Each bronchus splits into a number of smaller tubes, the **bronchioles**, which eventually end in a cluster of **alveoli** (singular: alveolus) (Figure 2.2). Gases are exchanged only in the alveoli – carbon dioxide goes out of the blood, and oxygen goes in. The respiratory system is protected by the ribs. The lungs are inflated and deflated using the **intercostal muscles** and the **diaphragm**.

▶ How air is breathed in and out

When the lungs expand, they suck air in; when they contract, they push air out again. Although they are elastic (springy), the

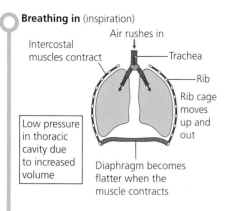

Breathing in (inspiration)

Intercostal muscles contract

Air rushes in

Trachea

Rib

Rib cage moves up and out

Low pressure in thoracic cavity due to increased volume

Diaphragm becomes flatter when the muscle contracts

Breathing out (expiration)

Air is pushed out

Intercostal muscles relax

Rib cage moves down and in

High pressure in thoracic cavity due to decreased volume

Diaphragm becomes dome-shaped when the muscle relaxes

Figure 2.3 Mechanism of breathing in and out

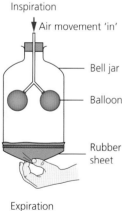

Inspiration

Air movement 'in'

Bell jar

Balloon

Rubber sheet

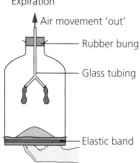

Expiration

Air movement 'out'

Rubber bung

Glass tubing

Elastic band

Figure 2.4 Bell jar model of the respiratory system

lungs are not muscular, so they cannot move on their own. The mechanism of breathing relies on the diaphragm (a sheet of muscle underneath the rib cage) and the rib cage itself, which is moved by the intercostal muscles between the ribs.

▶ When we breathe out, the intercostal muscles move the rib cage downwards and inwards, and the diaphragm moves upwards. This decreases the volume of the thorax and puts pressure on the lungs, so that the air in them is 'squeezed' out.

▶ Breathing in is the reverse process. The rib cage is moved upwards and outwards, and the diaphragm flattens. This increases the volume of the thorax, and the lungs, because they are elastic, naturally expand. The expansion of the lungs sucks air in through the trachea.

The movement of air into and out of the lungs is a result of differences in pressure between the air inside the lungs and the outside air. Gases move from areas of higher pressure to areas of lower pressure. The breathing mechanism creates a pressure inside the lungs that is lower than the outside air when breathing in, and a pressure that is higher than the outside air when breathing out.

The breathing mechanism is summarised in Figure 2.3. During **inspiration** (breathing in), both the intercostal muscles and the diaphragm are contracted, and during **expiration** (breathing out) all the muscles are relaxed.

Expiration is aided by the elasticity of the lungs. When they are not being stretched by air flowing in, they naturally recoil to help push air out.

We can model the respiratory mechanism and the respiratory system, as shown in Figure 2.4. The lungs are represented by the balloons, the rib cage by the bell jar, and the diaphragm by the rubber sheet.

Differences between inhaled and exhaled air

It is not true to say that we breathe in oxygen and we breathe out carbon dioxide. We breathe air in and out, but the composition of that air changes. The approximate figures are given in Table 2.1.

Table 2.1 Approximate composition of inspired and expired air

Gas	% in inspired air	% in expired air
Oxygen	21	16
Carbon dioxide	0.04	4
Nitrogen	79	79

Note that even in expired air there is a significant amount of oxygen, but the percentage is lower than in inspired air because some has been absorbed at the alveoli and replaced by carbon dioxide. The percentage of nitrogen remains unchanged because the body does not use that gas.

In addition, the expired air contains more water vapour than inspired air, because the surfaces of the alveoli are moist, and the air absorbs some water vapour while it is in the alveoli. As the internal temperature of the body at 37 °C is (usually) higher than that of the surrounding air, expired air also tends to be warmer than inhaled air.

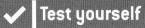

Test yourself

1 What is the name of the tubes which lead from the trachea into the lungs?
2 Where does gas exchange occur in the respiratory system?
3 Describe the movement of the rib cage and diaphragm during inspiration.
4 In the respiratory system, oxygen is exchanged for carbon dioxide. Where does this carbon dioxide come from?
5 Why does expired air contain more water vapour than inspired air?

▶ Digestion

Humans and all other animals get their energy from food. Food enters the gut, a tube that goes through the body. To be of any use, the food must move out of the gut and into the blood system, which then takes it to all parts of the body. The food we eat needs to be changed in two ways so that it can get out of the gut and into the blood system.

1 Large molecules in the food must be broken down into small molecules, which can be absorbed through the wall of the gut.
2 Insoluble molecules in the food must be changed into water-soluble ones, so they can dissolve in the blood and be transported around.

The process of digestion breaks down complex food molecules into small, soluble ones. All the chemical reactions involved are controlled by special chemicals called **enzymes**.

> **Key term**
>
> Enzyme Biological molecule which acts as a catalyst, speeding up a chemical reaction but not taking part in it.

What foods need digesting?

The complex food molecules in our diet fall into three categories:

▶ fats, which are broken down into glycerol and fatty acids
▶ proteins, which are broken down into amino acids
▶ carbohydrates, the main one being starch, which is an insoluble chain of glucose molecules and is broken down into single glucose molecules.

Glycerol, fatty acids, amino acids and glucose can all be absorbed readily into the blood. Energy is provided by glucose, glycerol and fatty acids, but amino acids are not normally respired. Instead, they are used as raw materials for making new proteins for growth.

▶ Enzymes

Enzymes are protein molecules that act as **catalysts**. A catalyst is something that speeds up a chemical reaction. It is unchanged by the reaction, but causes it to go faster. Here are some important facts about enzymes:

▶ Enzymes act as catalysts, speeding up chemical reactions.
▶ The enzyme is unchanged by the reaction it catalyses.

- Enzymes are specific, which means that a certain enzyme only catalyses one reaction or one type of reaction.
- Enzymes generally work better as temperature increases, but if the temperature gets too high they are destroyed (denatured). Every enzyme has an 'optimum' temperature at which it works best (human enzymes work best at body temperature, 37 °C).
- Different enzymes are denatured at different temperatures.
- Enzymes work best at a particular 'optimum pH' value, which is different for different enzymes.

How enzymes work

Enzymes work on chemicals called substrates. In order to catalyse a reaction, the enzyme has to 'lock together' with its substrate to form an **enzyme–substrate complex**. The shapes of the enzyme and substrate must match, so that they fit together like a lock and key. That is why enzymes are specific – they can only work with substances that fit into their **active site**.

This 'lock and key' model is shown in Figure 2.5.

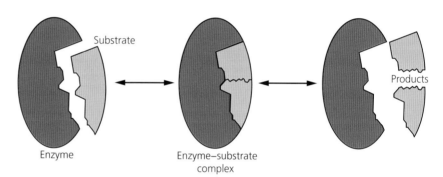

Substrate

Products

Enzyme

Enzyme–substrate complex

Figure 2.5 The 'lock and key' model of enzyme action. Note that in some reactions an enzyme catalyses the breakdown of a substrate into two or more products, while in others an enzyme causes two or more substrate molecules to join to make one product molecule

The effect of temperature and pH on enzymes

Warming an enzyme makes it work faster at first, because the enzyme and substrate molecules move around faster and so meet and join together more often. But at higher temperatures the enzyme stops working altogether.

The shape of the enzyme molecule is important. An enzyme is held in shape by chemical bonds. High temperatures and unsuitable pH conditions can break these bonds. This changes the shape so that the substrate molecule cannot join with it. The enzyme no longer works and is said to be denatured. Different enzymes denature at different temperatures. Some enzymes start to denature at about 40 °C, most denature at around 60 °C, and boiling denatures virtually all enzymes.

The effect of temperature on enzyme action is shown in Figure 2.6, and the effect of pH is shown in Figure 2.7.

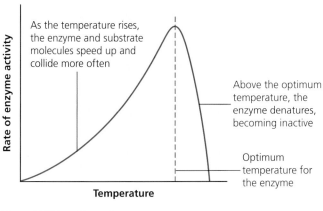

Figure 2.6 The effect of temperature on enzymes

As the temperature rises, the enzyme and substrate molecules speed up and collide more often

Above the optimum temperature, the enzyme denatures, becoming inactive

Optimum temperature for the enzyme

Rate of enzyme activity

Temperature

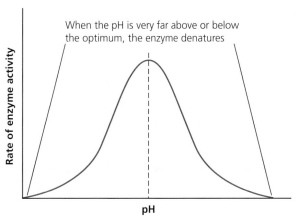

Figure 2.7 The effect of pH on enzymes

When the pH is very far above or below the optimum, the enzyme denatures

Rate of enzyme activity

pH

Test yourself

6 Why must large food molecules be broken down into smaller ones?
7 Why must insoluble molecules be changed into water-soluble ones?
8 What are the simple molecules that proteins are broken down into?
9 Enzymes are 'specific'. What does this mean?
10 Explain why high temperatures denature enzymes.

The human digestive system

Food is digested in the digestive system, sometimes called the gut. The useful products are absorbed into the blood as the food moves through the digestive system, and eventually the non-digestible parts are egested at the other end of the gut. Different parts of the gut have special functions. Figure 2.8 shows the structure and functions of the various parts of the digestive system. As well as the gut, the digestive system also includes some associated organs (the liver, gall bladder and pancreas). There are three processes, which occur in different parts of the digestive system:

1 **Digestion** – mainly in the mouth, stomach and small intestine
2 **Absorption** into the bloodstream – mainly in the small intestine (food) and large intestine (water)
3 **Egestion** – in the rectum (the lower part of the large intestine) and anus.

Key terms

Digestion The breakdown of food molecules into small, soluble molecules.

Absorption The movement of food molecules from the gut into the bloodstream.

Egestion The passage of undigested materials out of the body.

Peristalsis

To push food through your digestive system, waves of muscle contraction constantly move along the gut. These waves are called **peristalsis** (Figure 2.9).

When the circular muscles contract just behind the food, this squeezes the food forwards, rather like squeezing toothpaste from a tube.

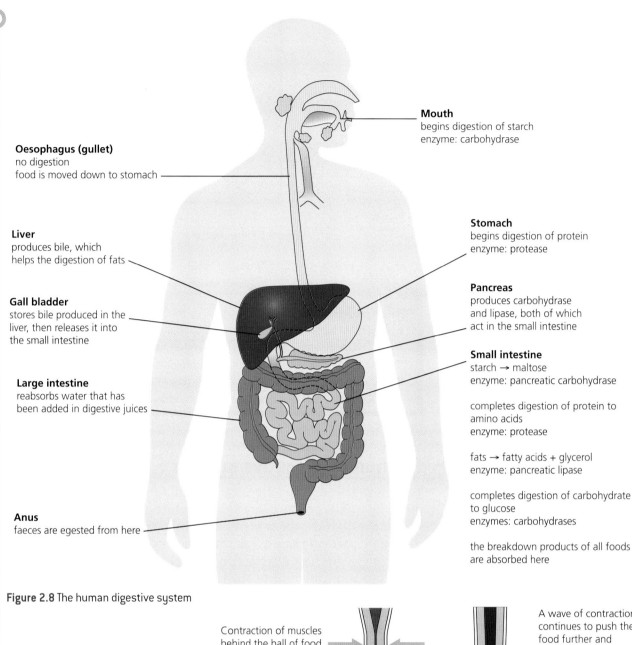

Mouth
begins digestion of starch
enzyme: carbohydrase

Oesophagus (gullet)
no digestion
food is moved down to stomach

Liver
produces bile, which
helps the digestion of fats

Gall bladder
stores bile produced in the
liver, then releases it into
the small intestine

Large intestine
reabsorbs water that has
been added in digestive juices

Anus
faeces are egested from here

Stomach
begins digestion of protein
enzyme: protease

Pancreas
produces carbohydrase
and lipase, both of which
act in the small intestine

Small intestine
starch → maltose
enzyme: pancreatic carbohydrase

completes digestion of protein to
amino acids
enzyme: protease

fats → fatty acids + glycerol
enzyme: pancreatic lipase

completes digestion of carbohydrate
to glucose
enzymes: carbohydrases

the breakdown products of all foods
are absorbed here

Figure 2.8 The human digestive system

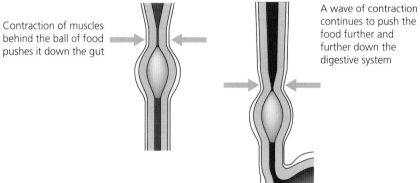

Contraction of muscles
behind the ball of food
pushes it down the gut

A wave of contraction
continues to push the
food further and
further down the
digestive system

Figure 2.9 Peristalsis in the gut

Bile

Bile is a liquid produced by the liver and stored in the gall bladder. When a meal containing fat is being digested, the gall bladder releases bile down the bile duct into the small intestine. Bile is not

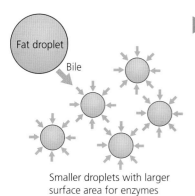

Smaller droplets with larger surface area for enzymes

Figure 2.10 The effect of bile on fats

an enzyme, but it helps the lipase enzyme in the small intestine to digest fats. Bile emulsifies fat, splitting it into small droplets and allowing a greater surface area for the lipase enzyme to work on (Figure 2.10).

Using the products of digestion

Once food substances have been digested into small, soluble chemicals, they can get through the wall of the gut and into your bloodstream, which will take them all around the body. This happens in the small intestine.

These substances are absorbed and used by the body:

▶ Glucose, formed from the breakdown of carbohydrates, is the main energy provider in the body, being broken down by respiration in the cells.
▶ Fatty acids and glycerol from fats also provide energy. Fats actually contain more energy per gram than glucose, but it can only be released slowly. For this reason, fats are useful as an energy store.
▶ Amino acids from proteins are re-assembled in the body into new proteins, to form many useful products or to be used for making new cells in growth.

Food tests

There are chemical tests for a number of the different food groups, including proteins and carbohydrates (with specific tests for starch and for glucose).

▶ Test for protein – A small volume of dilute sodium hydroxide solution is added to the test solution, then a roughly equal volume of **Biuret solution** is added. If protein is present, a purple colour is seen.
▶ Test for starch – When **iodine** solution is added to starch, the brown colour of the iodine turns to blue–black.
▶ Test for glucose – When a solution containing glucose is heated with blue **Benedict's solution**, a reddish-orange precipitate is formed. This is called the Benedict's test (Figure 2.11). The more glucose there is, the more precipitate is formed. As more and more precipitate is formed, the blue colour turns first to green, then orange, then to brick red.

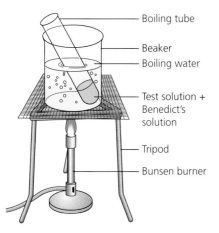

Boiling tube
Beaker
Boiling water
Test solution + Benedict's solution
Tripod
Bunsen burner

Figure 2.11 The Benedict's test

✔ Test yourself

11 In what parts of the digestive system does some sort of absorption take place?
12 What name is given to the process which moves food along the gut?
13 How does bile help digestion?
14 Why is it important for the body to take in amino acids?
15 A solution is tested using the Benedict's test and it goes brick red. What does this tell you about the solution?

Key term

Selectively permeable membrane

A membrane which allows some substances to pass through but not others, sometimes called a semi-permeable or partially permeable membrane.

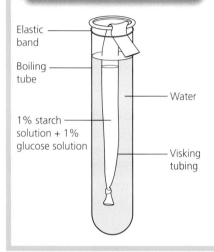

Elastic band

Boiling tube

Water

1% starch solution + 1% glucose solution

Visking tubing

Using a model gut

Visking tubing is a non-living material which acts as a **selectively permeable membrane**, like the membranes lining the gut. It lets small particles through but not large particles. Visking tubing can be used as a 'model gut'.

A student set up an experiment using Visking tubing as a model gut. The setup is shown in Figure 2.12.

Starch molecules are large and glucose molecules are small. The experiment was left for one hour. The liquids inside and outside the Visking tubing were tested using the iodine test and the Benedict's test.

1. Describe the results you would expect from the iodine and Benedict's tests.
2. Give reasons for your answers.
3. In what ways is Visking tubing (a) an accurate model of the human gut and (b) an inaccurate model of the human gut?

Figure 2.12 Apparatus used to model absorption in the gut

⚙ | Specified practical

Investigation into the factors affecting enzyme action

A student investigated the effect of temperature on an enzyme. Iodine is an indicator that turns blue–black when starch is present but is otherwise brown.

In this investigation, as the enzyme amylase breaks down the starch into sugar, the blue–black solution of starch and iodine changes to brown.

Procedure

1. $10\,cm^3$ of 1% starch solution were placed into a test tube.
2. $2\,cm^3$ of 10% amylase solution were placed into a second test tube.
3. Both tubes were placed in a water bath set at 20 °C for 3 minutes.

4. A drop of iodine was placed into separate wells of a spotting tile.
5. The test tubes were removed from the water bath, the amylase was added to the starch solution and the stopwatch was started.
6. Immediately, one drop of the mixture was added to the first drop of iodine, and the colour of the solution was recorded.
7. Step 6 was repeated every minute for five minutes.
8. Steps 1–7 were repeated at 30 °C, 40 °C, 50 °C and 60 °C.

Here are the student's results showing the colours of the solutions in the spotting tile:

Temperature (°C)	Time from beginning of experiment (mins)					
	0	1	2	3	4	5
20	Blue–black	Blue–black	Blue–black	Blue–black	Brown	Brown
30	Blue–black	Blue–black	Blue–black	Brown	Brown	Brown
40	Blue–black	Brown	Brown	Brown	Brown	Brown
50	Blue–black	Blue–black	Brown	Brown	Brown	Brown
60	Blue–black	Blue–black	Blue–black	Blue–black	Blue–black	Blue–black

Analysing the results

1 What are the **control variables** in this experiment?
2 Which temperature appears to be the optimum temperature for amylase enzyme?
3 Suggest what the student could do to increase confidence in the answer to question 2?
4 Explain the results at 60 °C.
5 What changes could you make to the procedure to test for the effect of pH on the activity of amylase?

Chapter summary

- The purpose of the respiratory system is to provide oxygen and remove carbon dioxide.
- The respiratory system consists of the nasal cavity, trachea, bronchi, bronchioles, alveoli, lungs, diaphragm, ribs and intercostal muscles.
- Air is breathed in and out as movements of the intercostal muscles and the diaphragm cause pressure and volume changes, so air is sucked in or forced out of the lungs.
- Inspired air has a different composition to expired air.
- Digestion is the process whereby large molecules are broken down into smaller molecules so they can be absorbed for use by body cells.
- Enzymes are proteins made by living cells that speed up or catalyse the rate of chemical reactions within the cells; specific enzymes are used for each reaction.
- Each enzyme has an optimum temperature and pH.
- Boiling destroys (denatures) enzymes.
- Enzyme activity involves molecular collisions; this is explained by the 'lock and key' model of enzyme action and formation of the enzyme–substrate complex at the active site.
- Soluble food substances are absorbed through the wall of the small intestine and eventually into the bloodstream.

- Visking tubing can act as a 'model gut', but the model has limitations.
- Fats are made up of fatty acids and glycerol, proteins are made up of amino acids, and starch is made up of an insoluble chain of glucose molecules.
- Fats, proteins and carbohydrates are broken down during digestion into soluble substances so that they can be absorbed.
- The structure of the human digestive system includes the mouth, oesophagus (gullet), stomach, liver, gall bladder, bile duct, pancreas, small intestine, large intestine and anus.
- Each of the following organs has a role in digestion: mouth, stomach, pancreas, small intestine, large intestine.
- Peristalsis is a process whereby food is moved along the digestive tract.
- Bile is secreted by the liver and stored in the gall bladder and it plays a role in the breakdown of fats.
- Fatty acids and glycerol from fats, and glucose from carbohydrate, provide energy while amino acids from digested proteins are needed to build proteins in the body.
- Food tests can test for the presence of starch using iodine solution, glucose using Benedict's reagent and protein using Biuret solution.

Practice exam questions

1 *Valonia ventricosa* is an unusual single-celled organism which lives in the seas of tropical and subtropical areas. It lives in shallow depths (80 m or less). The single cell is large, up to around 5 cm long. The cell has a cellulose cell wall, a vacuole and many nuclei and chloroplasts. It attaches to rocks by small hair-like structures called rhizoids. Its large size makes it easy to study and scientists have measured the concentrations of ions in the vacuole and the surrounding sea water. The results for some ions are shown below.

Ion	Concentration	
	Cell vacuole	Sea water
Potassium	0.5	0.01
Calcium	0.002	0.01
Sodium	0.1	0.5

a) State three features of *Valonia* that are found in plant cells. [3]

b) State two features of *Valonia* that are different from a normal plant cell. [2]

c) Look at the concentration data in the table. Suggest, with reasons, how each of the ions enters the cell (i.e. by diffusion or by active transport). [4]

d) Which of the ions shows the biggest difference in concentration between the sea water and the cell vacuole. [1]

2 Two men, Hefin and Carwyn, run on a treadmill for 8 minutes, and their breathing rates are monitored. Hefin rarely does any strenuous exercise, but Carwyn trains regularly. Their breathing rates before, during and after this exercise are shown in the graph below.

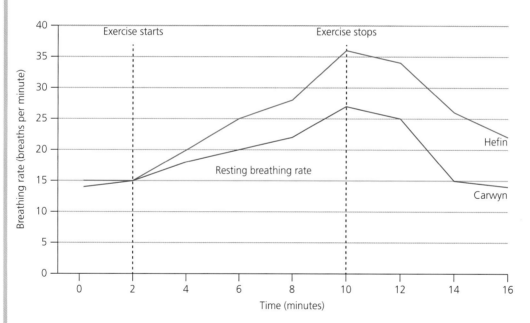

a) State two differences between the data for Hefin and Carwyn. [2]

b) Why is it important to take the resting breathing rate for both men before starting the exercise? [1]

c) What is the difference between Hefin and Carwyn's breathing rates at the end of the exercise period? [1]

d) Why does exercise increase breathing rate? [4]

e) It is likely that both Hefin and Carwyn's leg muscles switched to anaerobic respiration during this exercise. Why can this not continue for a long period? [2]

3 a) Identify the labelled structures shown in the diagram of the human respiratory system. [5]

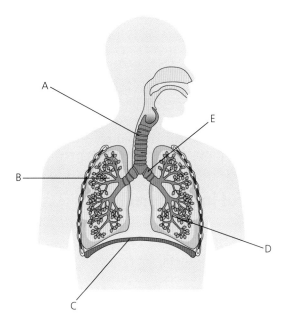

b) Name the structures in the respiratory system where gas exchange occurs. [1]

c) Indicate whether the statements that follow are true or false. [5]

i) Inspired air contains more water vapour than expired air.

ii) The diaphragm is made of muscle.

iii) During expiration, the diaphragm moves downwards.

iv) The volume of the thorax changes during breathing.

v) During inspiration, the pressure in the lungs is lower than that of the outside air.

4 An experiment was set up to investigate the action of amylase enzyme. Amylase enzyme catalyses the breakdown of starch into the sugar maltose. The apparatus was set up as shown in the diagram.

The experiment was left for 1 hour at room temperature (22 °C). After that time iodine solution was added to each tube.

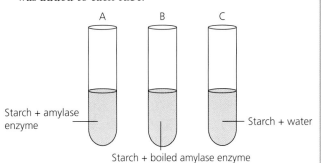

Starch + amylase enzyme

Starch + boiled amylase enzyme

Starch + water

The results were as follows:

Test tube	Colour when iodine added
A	Brown
B	Blue–black
C	Blue–black

a) Explain the different results in test tubes A and B. [4]

b) What variables need to be controlled in this experiment? [4]

c) Explain the purpose of test tube C. [1]

d) A student suggested that it would have been better to carry out the experiment at 30 °C. What changes would you expect to see in the outcome if this was done? [2]

e) Bread contains a lot of starch but also a small amount of glucose. When you eat bread, the starch needs to be digested but the glucose does not. Explain why this is the case.

3 Underpinning energy concepts

The use and transfer of energy is important in our daily lives. The most important **energy transfers** involve electricity. About 73% of electricity generated in Wales is from burning fossil fuels, which produces carbon dioxide gas, a greenhouse gas that contributes to global warming. Using electricity more efficiently will help us all.

▶ Why is electricity useful in modern life?

- ▷ It is easy to use electricity to transfer energy into useful energy stores.
- ▷ Electric current travels well through metal wires, so is easy to move over long distances from where it is generated to where it is needed.
- ▷ Electricity is easy to generate from stores of energy such as the chemical energy stored inside batteries.

✔ Test yourself

1 How much of electricity generated in Wales is from fossil fuels?
2 What are the three reasons why electricity is so useful to us?
3 State the energy form that is supplied to a power station and the useful energy output.

▶ Energy transfers and heating

Inside a power station, the transfer of energy is driven by heating. Energy moves from a place with a high temperature to a place with a low temperature.

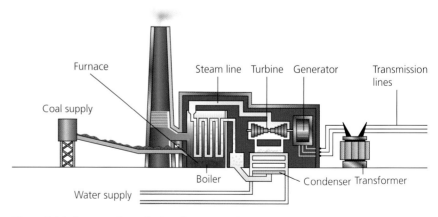

Figure 3.1 A diagram of a typical coal power station

In Figure 3.1, coal is burnt in the furnace at high temperature. The energy creates large amounts of steam in the boiler, which is at a lower temperature than the furnace. The steam moves to the even colder turbine, turning the generator and generating electricity. The steam is converted back to water in the condenser, the coldest part of the system at the lowest temperature.

Electrical power

'Power', measured in watts, W, is the rate at which a device transfers energy from one store into others – how much energy (in joules, J) the device can transfer per second (s):

$$\text{power} = \frac{\text{energy transfer}}{\text{time}}$$

Electrical power is the rate at which an electrical device changes stores of energy into electricity. It is also the rate at which a device, such as a kettle, transfers electricity into useful forms of energy.

The electrical power of a device can be calculated by multiplying the voltage (in volts, V) and the current (in amps, A) of the device together:

$$\text{electrical power} = \text{voltage} \times \text{current}$$

> **Key term**
>
> **Power** The rate at which a device transfers energy from one form into other forms.

★ Worked examples

1 A small portable fan has a power of 15 W. It runs for 5 minutes (300 s). Calculate the energy transferred.
2 Calculate the power of a 12 V light bulb with a current of 0.5 A flowing through it.
3 A 230 V lawn mower has a power of 2000 W. Calculate the current flowing through the lawn mower.

Answers

1 energy transfer = power × time = 15 W × 300 s = 4500 J
2 electrical power = voltage × current = 12 V × 0.5 A = 6 W
3 electrical power = voltage × current ⇒

$$\text{current} = \frac{\text{electrical power}}{\text{voltage}} = \frac{2000\,\text{W}}{230\,\text{V}} = 8.7\,\text{A}$$

Energy efficiency

When a mobile phone is working, some energy is wasted by heating causing the battery and the phone to warm up. Mobile phone batteries are very good at doing their job: for every 100 J of chemical energy stored in the battery, 98 J is transferred into electricity and only 2 J is wasted. As the batteries transfer such a large amount of their stored chemical energy into useful electricity (and little is wasted), we say that they are very efficient.

Key term

Efficiency The ratio of energy (or power) usefully transferred/total energy (or power) supplied, which is normally expressed as a percentage.

The **efficiency** of a device or a process is normally expressed as a percentage (%). A gas power station for example, is about 30% efficient. A device that converts all its available input energy into useful output energy is 100% efficient. The more efficient a device is, the less energy is wasted.

Efficiency is calculated using the following formula:

$$\% \text{ efficiency} = \frac{\text{energy } [\text{or power}] \text{ usefully transferred}}{\text{total energy } [\text{or power}] \text{ supplied}} \times 100$$

★ Worked examples

1 The battery in a mobile phone holds 18 000 J of energy. If the battery transfers 16 000 J usefully and 2000 J is wasted, what is the efficiency of the battery?
2 A power station transfers electricity to the National Grid with a power of 60 MW. A total power of 200 MW is supplied by burning coal. What is the efficiency of the power station?
3 A solar panel is rated as 30% efficient. The power usefully transferred by the panel is 180 W. What is the total power supplied by sunlight to the panel?

Answers

1 Total energy supplied = 18 000 J
Energy usefully transferred = 16 000 J

$$\% \text{ efficiency} = \frac{\text{energy usefully transferred}}{\text{total energy supplied}} \times 100$$

$$\% \text{ efficiency} = \frac{16\,000 \text{ J}}{18\,000 \text{ J}} \times 100 = 89\,\%$$

2 Total power supplied = 200 MW
Power usefully transferred = 60 MW

$$\% \text{ efficiency} = \frac{\text{power usefully transferred}}{\text{total power supplied}} \times 100$$

$$\% \text{ efficiency} = \frac{60 \text{ MW}}{200 \text{ MW}} \times 100 = 30\,\%$$

3 Efficiency = 30%
Power usefully transferred = 180 W

$$\% \text{ efficiency} = \frac{\text{power usefully transferred}}{\text{total power supplied}} \times 100$$

Rearranged:

$$\text{total power supplied} = \frac{\text{power usefully transferred}}{\% \text{ efficiency}} \times 100$$

$$\text{total power supplied} = \frac{180 \text{ W}}{30\,\%} \times 100 = 600 \text{ W}$$

✓ Test yourself

4 Bulbs A, B and C in the table below all give out the same light output.

Type of bulb	Power usefully transferred (W)	Total power supplied (W)	% Efficiency
A	1.5	50	
B	1.5	15	
C	1.5	2.0	

a) Calculate the efficiency of each type of bulb.
b) Which type of bulb would you fit in your bedroom? Explain your answer.

c) Bulb C is used in a torch for 120 s. Calculate the useful energy transferred.
d) A fourth bulb, D, has a voltage of 4.5 V with a current of 3.0 A running through it. Calculate the power of bulb D.
5 The power usefully transferred from a large wind turbine is 0.50 MW. The total power supplied by the wind is 0.75 MW. Calculate:
a) the efficiency of the wind turbine.
b) the amount of wind power wasted by the turbine.

Modern living and energy

20

▶ Sankey diagrams

Sankey diagrams show energy (or power) transfers as a diagram. They are drawn to scale and show the relative or percentage amount of energy transferred. The widths of the bars on the Sankey diagram show the amount of energy involved, and so the efficiency is shown by the width of the useful energy bar compared with the width of the total input energy bar.

★ Worked example

Draw a Sankey diagram for an energy-efficient light bulb. Every second, 10 J of energy is transferred to the bulb. 2 J is output as useful light, 8 J is wasted.

Answer

Draw this Sankey diagram so that the input bar is 10 units wide, the useful bar is 2 units wide, and the wasted energy bar is 8 units wide. Usually, the useful energy transfer runs along the top of the diagram (as a straight bar) and the wasted energies curve away below. Of the 10 J of energy input, only 2 J is output as useful light, so 'low-energy' light bulbs are only 20% efficient!

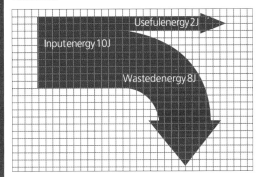

Figure 3.2 A Sankey diagram for an energy-efficient bulb

✔ Test yourself

6 An LED light bulb is 80% efficient. A total of 100 J of energy is supplied to the LED. 80 J is transferred usefully and 20% is wasted. Draw a Sankey diagram for this bulb.

7 Figure 3.3 shows the Sankey diagram for an electric toothbrush. 55 J of energy is usefully transferred.

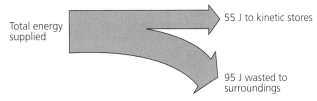

Total energy supplied

55 J to kinetic stores

95 J wasted to surroundings

Figure 3.3 The Sankey diagram for an electric toothbrush

a) Calculate the total energy supplied to the electric toothbrush.
b) Calculate the % efficiency of the electric toothbrush.

Worked example

A student performs an energy efficiency investigation on a kettle. She pours 1 kg of water into the kettle and connects the kettle to the mains via an energy efficiency plug. The water is initially at a temperature of 15 °C and boils at 100 °C. The energy usefully transferred to the water from the kettle is given by the equation:

energy usefully transferred (J) = mass of water (kg) × 4200
× temperature change (°C)

The energy efficiency plug measures the total energy supplied as 446 250 J.

a) Calculate the temperature change of the water.
b) Calculate the energy usefully supplied by the kettle.
c) Calculate the % efficiency of the kettle.

Answers

a) temperature change = final temperature − initial temperature
= 100 °C − 15 °C = 85 °C

b) energy usefully transferred = mass of water × 4200 × temperature change = 1 kg × 4200 × 85 °C = 357 000 J

c) % efficiency of the kettle = $\dfrac{\text{energy usefully transferred}}{\text{total energy supplied}} \times 100$

$= \dfrac{357\,000\ \text{J}}{446\,250\ \text{J}} \times 100 = 80\%$

Specified practical

Investigation of the efficiency of energy transfer in electrical contexts

A student modelled the efficiency of the energy transferred from a kettle using a resistor and a beaker of water.

A diagram of the student's apparatus is shown in Figure 3.4.

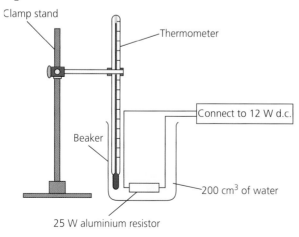

Figure 3.4 Diagram of the apparatus used to measure the efficiency of energy transfers from a heating element

Procedure

1 200 cm³ of water was added to the beaker and the 12 V d.c. power supply was turned on.
2 The temperature of the water was measured and recorded every 60 seconds for 600 seconds. The student collected the following results:

Time /seconds	Temperature /°C
0	20
60	21
120	22
180	23
240	24
300	25
360	26
420	28
480	30
540	32
600	34

1 Draw a graph of time (x-axis) against temperature (y-axis). Start your y-axis at 15 °C and finish at 35 °C.
2 Calculate the temperature rise of the water.
3 Calculate the total energy supplied to the water by the resistor using the equation:

$$\text{energy transfer (J)} = \text{power (W)} \times \text{time (s)}$$

4 Calculate the energy usefully transferred to the water using the equation:

$$\begin{aligned}\text{energy usefully}\\\text{transferred (J)}\end{aligned} = \begin{aligned}\text{temperature rise (°C)}\\\times 840\end{aligned}$$

5 Calculate the % efficiency of the resistor using the equation:

$$\text{\% efficiency of the resistor} = \frac{\text{energy usefully transferred}}{\text{total energy supplied}} \times 100$$

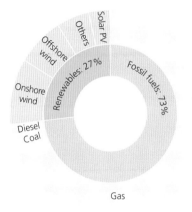

Figure 3.5 Electricity generation in Wales by fuel type in 2019

Global warming The gradual increase in overall average global atmospheric temperature.

Renewable Energy sources produced by the action of the Sun and not used up when working.

Sustainability Using renewable energy and using that energy very efficiently.

Carbon footprint The equivalent mass of carbon dioxide gas produced when an energy source generates electricity.

▶ Sustainable electricity and carbon footprint

Figure 3.5 shows that 73% of the electricity in Wales is generated by power stations burning fossil fuels: coal, oil (diesel) and natural gas. Fossil fuels contain carbon, and when they are burned the carbon reacts with oxygen forming carbon dioxide gas, which contributes to **global warming**.

In 2019 27% of the electricity in Wales was generated from '**low-carbon**' sustainable sources, such as wind or solar power. These sources are sustainable because they are **renewable** and are produced by the action of the Sun.

Sustainable electricity generation produces minimal carbon dioxide and so makes little contribution to global warming and does not put at risk the future ability of people to meet their own energy needs.

Reducing global warming reduces the risks associated with potential human disasters such as widespread flooding and drought.

Measuring carbon footprint

The **carbon footprint** of an energy source is the equivalent mass of carbon dioxide gas produced when the energy source generates electricity. The carbon footprint of coal is very high because a large mass of carbon dioxide gas is produced when it burns. The carbon footprint of wind energy is very low, because the only fossil fuels involved are during the manufacture, installation and de-commissioning of the wind turbine.

The measurement of carbon footprint is given in terms of the mass equivalent of carbon dioxide (**kgCO2eq**). This is determined by the mass of gas produced and its global warming potential:

$$\text{kgCO2eq} = (\text{mass of gas}) \times (\text{global warming potential of the gas})$$

The global warming potential of three common gases is given in Table 3.1.

Table 3.1 The global warming potential of three common gases

Common gas	Global warming potential
carbon dioxide, CO_2	1
methane, CH_4	21
nitrous oxide, N_2O	298

★ Worked example

When a wagon load of coal is burned in a power station, 97 kg of carbon dioxide and 3 kg of nitrous oxide are produced. Calculate the carbon footprint of burning the wagon load of coal in kgCO2eq.

Answer

kgCO2eq of carbon dioxide = mass of carbon dioxide × global warming potential of carbon dioxide

kgCO2eq of carbon dioxide = 97 kg × 1 = 97

kgCO2eq of nitrous oxide = mass of nitrous oxide × global warming potential of nitrous oxide

kgCO2eq of nitrous oxide = 3 kg × 298 = 894

Total kgCO2eq = 97 + 894 = 991 kgCO2eq

✔ Test yourself

8 Use Figure 3.5 to estimate the percentage of electricity that is generated in Wales from wind energy.
9 A research scientist collects a sample of permafrost (frozen ground) during an expedition to Greenland. When she melts the permafrost, she finds that the sample contains 20 kg of methane gas. Calculate the carbon footprint of this sample.

⬇ Chapter summary

- Temperature differences lead to the transfer of energy by heating.
- Sankey diagrams can be used to show energy transfers.
- The efficiency of an energy transfer can be calculated from the equation:

$$\%\text{efficiency} = \frac{\text{energy} \left[\text{or power} \right] \text{ usefully transferred}}{\text{total energy} \left[\text{or power} \right] \text{ supplied}} \times 100$$

- power = voltage × current
- energy transfer = power × time
- Sustainability involves using renewable energy and then using that energy very efficiently.
- The carbon footprint of an energy source is the equivalent mass of carbon dioxide gas produced when the energy source generates electricity.
- The measurement of carbon footprint is given in terms of the mass equivalent of carbon dioxide (kgCO2eq) and is determined by the mass of gas produced and its global warming potential:

kgCO2eq = (mass of gas) × (global warming potential of the gas)

4 Generating electricity

The generation of electricity is fundamental to sustainability. Electricity is our most important source of energy, and we must ensure that the environmental impacts of generation are minimised while maintaining a safe and secure supply.

▶ The advantages and disadvantages of different energy technologies

Generating electricity has advantages and disadvantages whatever the primary energy source. Table 4.1 compares the **renewable** forms (wind; solar; hydroelectric; wave and tidal; biofuels; and geothermal) to the **non-renewable** forms (coal; oil; gas; and nuclear).

Table 4.1 The advantages and disadvantages of generating electricity from different primary sources of energy

Primary source of energy	Advantages	Disadvantages
Fossil fuels (such as coal, oil and gas)	Large amounts of electricity generated cheaply. Reliable. Security of supply.	Dangerous soot particles released into the air (especially by coal). Large carbon footprint. Carbon dioxide gas produced which contributes to global warming. Sulfur dioxide gas produced, contributing to acid rain. Large amounts of fuel brought onto site, and waste (in the case of coal) must be removed from the site. Non-renewable forms of energy are only 35 to 55 % efficient.
Nuclear energy	Low carbon footprint – limited carbon dioxide gas emitted. Energy produced for long periods of time without the need for refuelling. Very reliable. Lots of energy produced.	Expensive to build and decommission. Radioactive waste needs to be stored securely for a very long time. Non-renewable. There is risk of terrorist attack. Danger of potential nuclear accident.
Wind energy	Renewable. No air pollution. Very low carbon footprint.	Windy sites are remote – unsightly high-voltage power lines needed. Unreliable as only operate when it is windy. Many turbines needed. Can be unsightly and noisy to local residents.
Solar power	Renewable. Very low carbon footprint. Predictable and reliable (in daytime). Cheap to install. Solar panels can be retro-fitted to buildings. Easy to install in areas with large populations.	Does not generate electricity at night. Large-scale solar power stations use up a lot of land. Large areas of solar panels are needed to generate large amounts of electricity.
Hydroelectric power (HEP)	Renewable. No air pollution. Large amounts of electricity generated. Instant start-up times.	Expensive to build. Destroy habitats. HEP sites are remote – unsightly high-voltage power lines needed. Unreliable during a drought.
Wave and tidal energy	Renewable. Tides are very predictable and reliable. Large amounts of electricity generated from tidal. Non-polluting. Very low carbon footprint. Can be turned on and off very quickly.	Wave energy is unreliable (only works when there are suitable waves). Large numbers of wave generators needed to generate significant amounts of energy. Tidal energy barrages destroy habitats.

Primary source of energy	Advantages	Disadvantages
Biofuels (such as animal waste, wood and fast-growing crops)	Renewable. Large-scale biofuel power stations could be built, generating large amounts of electricity.	Large areas of land needed for fast-growing crops, or large amounts of animal waste needed, which would have to be transported cleanly. Carbon neutral, but carbon dioxide still emitted into the atmosphere. Half the carbon footprint of fossil fuels. Unsightly.

▶ Comparing two forms of electricity generation

When considering which types of electricity generation methods to build in a particular location many competing factors need to be considered:

- ▶ sustainability
- ▶ carbon footprint
- ▶ costs, including building and decommissioning costs, and cost of electricity generated
- ▶ reliability (or security)
- ▶ impact on the environment, including emission of global warming and acid rain gases, and visual and noise pollution.

Figure 4.1 North Hoyle offshore wind farm

Table 4.2 below compares Indian Queens gas turbine power station in Cornwall (a fossil fuel power station), to the North Hoyle offshore wind farm in North Wales (a renewable energy facility).

Table 4.2 Comparing methods of electricity generation

	Indian Queens gas turbine power station	North Hoyle offshore wind farm
Description	A single large gas turbine jet engine that is only used at peak demand time (it is only used for about 450 hours per year).	A 30-turbine wind farm situated in one of the windiest places in the UK.
Primary energy source	Liquid kerosine (jet fuel) or diesel (both fossil fuels).	Wind
Reliability	The power station can reliably produce 140 MW of electricity at very short notice (8 minutes) and can run at full power for 24 hours before shutting down.	Depends on wind strength, but North Hoyle generates at full power for 35% of the time.
Environmental issues	Fossil fuels are burned producing large quantities of carbon dioxide, nitrous oxide and sulfur dioxide gas. These go into the atmosphere contributing towards global warming and acid rain. The power station is situated on the Goss Moor, a government nature site and can be seen for miles. The gas turbine produces a lot of noise when in operation.	North Hoyle produces no gases that contribute to global warming or acid rain. The large turbines can be clearly seen from along the local coastline, and act as a hazard to the local bird populations. Wind turbines produce some noise when operating but North Hoyle is 5 miles offshore.
Electricity power output	140 MW	60 MW
Total power input	425 MW	Depends on wind strength (but typically 133 MW)
% efficiency	33%	Depends on wind strength (but typically 45%)
Carbon footprint per year (kgCO2eq)	57 000	0
Sustainability	Very low – does not use renewable energy sources and is only 33% efficient.	High – uses a renewable energy source (the wind) and is typically 45% efficient.
Set-up cost	£60 million	£80 million
Cost price of electricity	£300 per MWh	£77 per MWh
Projected lifetime	30 years	25 years
Projected decommissioning cost	£4 million	£24 million

✔ Test yourself

1 What is the useful power output by Indian Queens power station?
2 Suggest a reason why the carbon footprint of North Hoyle offshore windfarm is 0 kgCO2eq.
3 Suggest a reason why the total power output of North Hoyle offshore windfarm is not given as a constant value.
4 Compare the % efficiencies of these two electricity generating stations.

5 The cost price of electricity produced by other large fossil fuel power stations in the UK is about £60 per MWh. Suggest a reason why the electricity produced by Indian Queens power station is so much higher.
6 Suggest two reasons why the projected decommissioning costs of North Hoyle offshore wind farm are so much higher than the projected decommissioning costs of Indian Queens power station.

▶ Could our homes be used as micro power stations?

Individual homes, schools, businesses, and government buildings could be fitted with solar panels and micro wind turbines. Combine this with a programme of building insulation, and we would reduce demand and generate a considerable amount of electricity for our domestic and commercial consumption. However, it's not always daylight, the wind does not always blow, and large-scale industry needs guaranteed access to large amounts of power that could not be delivered by local micro-generation. What we need is a mixture of large-scale power stations and renewable sources of energy, improved insulation, and more energy-efficient devices.

A micro-generation system, such as a small wind turbine or solar panels, is a **cost-effective** option for some households, particularly new build houses. The cost-effectiveness depends on the **payback time** for the system. This depends upon the cost of installing the system and its savings in fuel costs.

$$\text{payback time (days)} = \frac{\text{cost of installation (£)}}{\text{savings in fuel costs (£ per day)}}$$

The shorter the payback time, the bigger the cost-effectiveness of the system.

Figure 4.2 A house fitted with a micro-turbine and solar panel

★ Worked example

A houseowner is quoted £8000 to install solar panels on her house, giving a saving of £1.30 per day. Calculate the payback time, in years, of buying the panels. 1 year = 365 days.

Answer

$$\text{payback time (days)} = \frac{\text{cost of installation (£)}}{\text{savings in fuel costs (£ per day)}}$$

$$\text{payback time} = \frac{£8000}{£1.30 \text{ per day}} = 6154 \text{ days} = 16.9 \text{ years}$$

The operational lifetimes and useful output powers of several renewable micro-generation devices are shown in Table 4.3. All of these devices can be fitted to houses. They all generate renewable energy, but they also have their disadvantages.

Table 4.3 Comparing the operational lifetimes and the useful power produced by some renewable energy devices to a gas power station

Energy source	Cost per kW	Operational lifetime (years)	Useful power produced (kW)	Disadvantages
Solar panel	£1200	40	0.350 kW per panel	Large roof areas needed. South facing roofs best.
Hydrogen fuel cell stack	£5000 to buy £75 per kW to run	10	1 kW	Good access to liquid hydrogen needed.
Micro wind turbines	£7000	20	0.15 kW	Only suitable in windy places.
Micro water turbine	£10 000	50	0.10 kW	Only suitable for houses close to running water
Gas power station	£800	40	520 000 kW	Burns non-renewable fossil fuel, which releases carbon dioxide gas.

✔ Test yourself

7 Which renewable energy device is the cheapest way to provide an electrical power of 1 kW?

8 A householder wants to install a 4 kW solar panel system. Calculate the cost of this system.

9 Explain why a hydrogen fuel cell has an ongoing running cost.

10 The time (in hours) needed to generate an amount of electrical energy (in kWh), with a given electrical power (in kW) is given by the equation:

$$\text{time (hours)} = \frac{\text{amount of electrical energy (kWh)}}{\text{power (kW)}}$$

A homeowner installs the combined renewable energy system shown in Figure 4.2. This consists of a 0.15 kW micro wind turbine and two solar panels with a **combined** power output of 0.85 kW. During the day, the home requires 8 kWh of electrical energy supplied.

a) Use the equation above to calculate the total time (in hours) that the combined system needs to be generating electricity.

b) Explain why this renewable energy system is unlikely to meet all the energy requirements of a typical household.

▶ How do we move electricity around?

The patterns in the UK national consumption of electricity are shown in Figure 4.3.

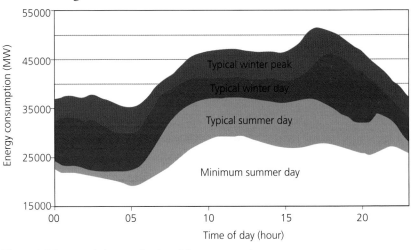

Figure 4.3 Seasonal changes in electricity consumption

The National Grid, shown in Figure 4.4, is responsible for matching the annual and daily consumption of electricity with production in the UK. The National Control Centre in Wokingham is responsible for balancing these two competing processes.

The UK's electricity is generated in a vast network of power stations, wind farms, and hydroelectric plants, all connected together and to us by a network of cables. National Grid operators are constantly predicting the demand for electricity in 30-minute blocks, and then directing power generators to supply the relevant amount. Some power stations are constantly producing electricity. Others, like Indian Queens power station, are only needed at peak times. The combined effect of this complex system is to match supply to demand. Without this system, the security of our electricity supply would be compromised, and we would spend significant amounts of time without power.

Figure 4.4 The National Grid in Wales and England

▶ The National Grid transmission system

When electric current flows through a wire, the wire heats up. If electricity was transmitted around the country at low voltage and high current, the amount of energy wasted by heating would be colossal and the price of electricity would be excessive. Therefore, electricity is transmitted around the country at very high voltage but very low current to minimise energy loss by heating.

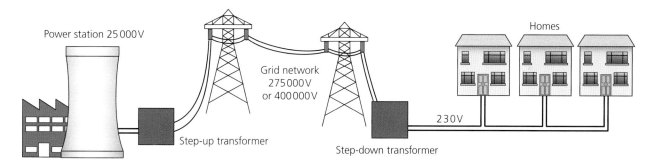

Figure 4.5 The National Grid transmission system

High voltage/low current supplies are very dangerous and cannot be used by household devices. So, the National Grid's electricity needs to be changed by a **transformer** before it enters our homes. **Step-up transformers** are found at power stations. They transform the electricity into high voltage/low current to allow it to be transmitted over long distances. **Step-down transformers** are found near our homes. They convert the electricity into safer low voltage/high current. In total, the National Grid is about 92% efficient.

11 Why is electricity transmitted around the National Grid at very high voltage?
12 Why do we not use high-voltage electrical devices at home?
13 What is the name of the device that changes the voltage and current of electricity?

Specified practical

Investigation of the factors affecting the output from a solar panel

A student carried out an experiment to investigate how the output voltage of a panel of solar cells varies with the intensity of the light radiation falling on it. She varied the light intensity by moving a light source further away from the panel.

A diagram of the apparatus that the student used is shown in Figure 4.6.

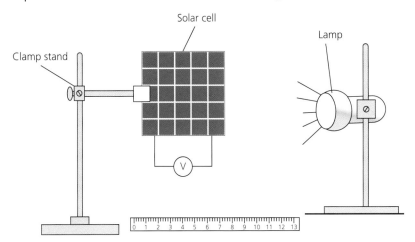

Figure 4.6 Diagram of the apparatus used to measure the voltage output of a solar panel with varying light intensity

Procedure

1 The lamp was placed 20 cm away from the solar panel and turned on.
2 The output voltage of the solar panel was measured and recorded.
3 Step 2 was repeated for distances of 40, 60, 80 and 100 cm away from the solar panel.
4 The experiment was repeated twice.

Results

Distance of lamp from the solar panel (cm)	Voltage output of the solar panel (V)			
	1st attempt	2nd attempt	3rd attempt	mean
20	2.4	2.0	2.4	2.4
40	1.3	1.3	1.3	1.3
60	0.7	0.8	0.6	
80	0.4	0.5	0.6	0.5
100	0.3	0.3	0.3	0.3

Analysing the results

1 Identify the anomalous data value in the table.
2 Suggest a reason why the data value chosen in Question 1 is anomalous.
3 Calculate the missing mean value.
4 Plot a graph of mean voltage output (y-axis) against distance (x-axis) and draw a suitable best-fit line.
5 Describe the pattern in the results.
6 Use your graph to estimate the voltage output of the solar panel when the lamp is 30 cm from the solar panel.

Chapter summary

- All forms of electricity generation have advantages and disadvantages.
- Renewable energy technologies include: hydroelectric, wind, waves, tidal, waste, crops, solar and wood.
- Non-renewable sources of power generation include the fossil fuels (coal, oil, and gas) and nuclear fuel.
- The cost effectiveness of introducing domestic solar and wind-energy equipment into a house is determined by the installation cost of the equipment and the fuel cost savings. The payback time is the duration (in years) before the savings start to outweigh the installation costs.
- When comparing different methods of power generation, the following factors all need to be considered: lifetime; useful power output; efficiency; reliability; carbon footprint; sustainability; environmental impact (including the atmospheric effects of carbon dioxide emission and acid rain, and pollution including visual and noise pollution); reliability; and costs.
- The national electricity distribution system (the National Grid), monitors and maintains a reliable energy supply that is capable of responding to varying demand.
- Electricity is transmitted efficiently across the country at high voltages, but low voltages are used at home because it is safer.
- Transformers are needed to change the voltage and current within the National Grid.

It is vital we use electricity carefully and efficiently, so that as little as possible is wasted. Approximately 30% of the energy used in the UK is used by households, and 40% by transport. If we use this energy effectively and efficiently, we can reduce the amount that we need to use, save money and live sustainably.

▶ Running electrical appliances

The running cost of an electrical device, such as a washing machine, depends on the electrical power of the device (normally given in kW) and the electricity tariff (the cost per unit of electrical energy). Remember, the electrical energy used by an appliance is calculated using the equation:

$$\text{energy transferred} = \text{power} \times \text{time}$$

When calculating the cost of domestic electricity, supply companies use units called **kilowatt hours** (kWh) or simply 'units'. 1 kWh is the amount of electrical energy consumed by a standard 1 bar (1 kW) electric fire in 1 hour. Units of electrical energy are calculated using the equation:

$$\text{units used (kWh)} = \text{power (kW)} \times \text{time (h)}$$

The cost of the electrical energy is calculated by multiplying the number of units by the cost per unit:

$$\text{total cost} = \text{units used} \times \text{cost per unit}$$

The cost per unit of electrical energy depends on the energy supply company and the type of electricity plan – in 2022 the cost of a unit of electricity in the UK was about 28 p.

> **Key term**
>
> **Kilowatt hour** The unit of electrical energy consumed.

★ | Worked examples

1 A 100 W lamp is left on for 10 minutes (600 s). How much electrical energy is transferred?

2 Suppose that you leave a 3 kW heater on in your room. You put it on at 8 am and forget about it until you get home at 4 pm. If a unit (1 kWh) costs 28 p, how much will it have cost to leave the heater on?

Answers

1 energy transferred = power × time
energy transferred = 100 W × 600 s = 60 000 J = 60 kJ

2 number of units used (kWh) = power (kW) × time (h)
   number of units = 3 kW × 8 h = 24 kWh (units)
   cost = number of units × cost per unit
   cost = 24 kWh × 28 p = 672 p = £6.72

Test yourself

1 Beth is worried. She left the heater in her room on from 7.00 am until 5.00 pm. It is a 3 kW heater.
   a) How many hours was it on for?
   b) How many units of electricity did it use?
   c) If the electricity cost 28 p per unit, how much did her mistake add to the family electricity bill?
2 Which of the following appliances costs most to run per day?
   a) 4 kW cooker on for 1 hour

   b) Six 1 kW flood light bulbs on for 4 hours
   c) a 1 kW washing machine on for 45 minutes (0.75 hours)
   d) a 0.045 kW Playstation on for 3 hours.
3 A rugby club replaces 24 × 2 kW floodlights with 12 × 0.1 kW LED floodlights. Both systems give out the same amount of light and are on for 8 hours per week. If electricity costs 28 p per unit, how much does the club save per week?

How much energy could you save?

Figure 5.1 shows an energy banding label for a household appliance, such as a washing machine, and gives important information about the energy efficiency and consumption of the device (as well as extra information such as its noise level).

A-rated devices are the most efficient, and G are the worst. The energy consumption is given in kWh per time interval (per 100 hours as in Figure 5.1, or per annum). When buying a new appliance, this label allows you to compare the running costs and sustainability of different models. The energy consumption value multiplied by the cost per kWh (or unit) will give you the total cost of running the appliance for the time interval stated on the label.

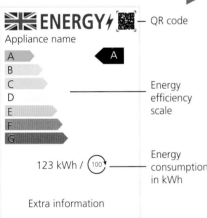

Figure 5.1 Energy banding label

> **Key term**
>
> Energy efficiency Rating of appliances is on an A–G comparison scale, where A-rated devices are more efficient in their use of energy than G-rated devices.

Reducing energy loss to thermal stores

All forms of thermal insulation work in the same way – they prevent energy moving from somewhere hot to somewhere cold. Reducing energy lost from a house to the thermal energy store of its surroundings reduces the amount of energy needed to heat the house. This saves money but also reduces the carbon footprint of the house; less electricity (or gas) is needed and so fewer fossil fuels are burned, reducing the amount of carbon dioxide escaping into the atmosphere.

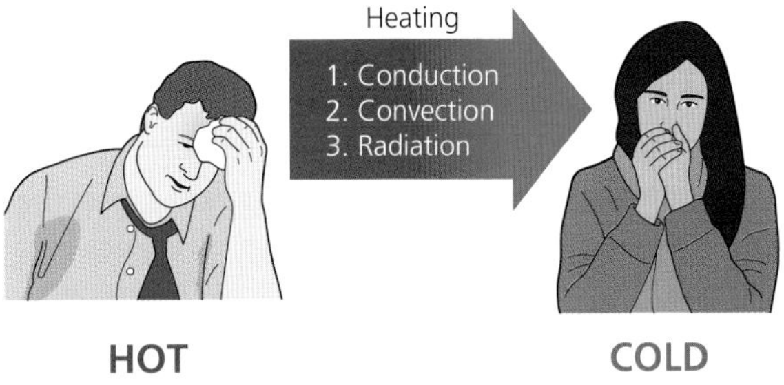

Figure 5.2 Energy transfer by heating

Insulation reduces energy losses by reducing the effect of the three mechanisms of heat transfer:

- convection
- conduction
- radiation

Reducing convection

Draught excluders work by reducing **convection currents** under a door or through the gaps in the frame, saving 10% to 20% of a household's heating costs. Convection currents occur as warm air rises and is replaced by colder air.

Draught excluders act as barriers between hot and cold places. The cold air outside a room cannot be sucked through the gap under a door as the hot air inside the room rises. The draught excluder prevents the colder air particles from entering the room.

Reducing conduction

Another cost-effective way to save on a heating bill is to install a thick layer of loft insulation. The materials used in loft insulation are poor **conductors** of energy by heating. Generally, the air inside rooms tends to be quite warm, and the air at the top of the room tends to be hottest because of convection currents. The air inside loft spaces tends to be quite cold and energy moves through the material of the ceiling, from hot (the room) to cold (the loft).

Energy is transferred through the material of the ceiling by vibrations of its particles. This process is called conduction. Conduction can happen in solids and liquids when the particles of matter vibrate. The hottest particles vibrate most and pass their vibration onto the (colder) particles next to them.

Energy does not transfer well by heating through non-metals. The materials used for loft insulation are generally made from non-metals, such as wool, glass or mineral fibres. The fibres also trap air, an insulator, between them. The thicker the layer of insulation, the better it works. Thicker insulation, however, is more expensive.

Double-glazed windows also reduce energy loss via conduction. There is a vacuum (or layer of air) between the panes of glass. There are no particles in a vacuum, so no conduction can occur. Air, being a poor conductor, acts in much the same way.

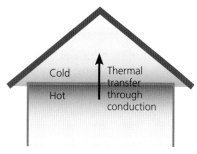

Figure 5.3 Energy transfer in a house with a loft

Reducing radiation

When new houses are built, the outside walls are made from a double layer of bricks sandwiching a thick layer of shiny, foil-coated polyurethane insulation panel. The shiny foil acts like a mirror and reflects infrared radiation, emitted by the warmer inside wall, back into the room and reduces energy loss via radiation.

Older houses without cavity wall insulation can have insulating foam pumped into the cavity between the layers of bricks. This reduces convection currents in the cavity wall and so reduces the amount of energy radiated away by the outer layer of bricks.

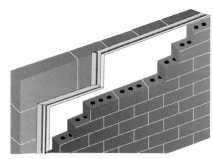

Figure 5.4 Cavity wall insulation panels

⚙ Specified practical

Investigation of the methods of heat transfer

A student performed experiments to investigate the methods of heat transfer. Each experiment looked at a different method.

Diagrams of the apparatus that the student used are shown in Figures 5.5, 5.6 and 5.7.

Convection experiment procedure

1 The apparatus was set up as shown in Figure 5.5.
2 A single crystal of potassium manganate(VII) was dropped into the beaker, close to one side.
3 The crystal was heated from below with a small, gentle Bunsen burner flame and the observations were recorded.

Radiation experiment procedure

1 The apparatus was set up as shown in Figure 5.6.
2 Adhesive tape was used to fix a 2 cm strip of black paper to the bulb of one of the thermometers, and a 2 cm strip of aluminium foil to the bulb of the other thermometer.
3 The temperature of each thermometer was measured and recorded.
4 The filament lamp was turned on and the temperature of each thermometer after 10 minutes was measured and recorded.

Conduction experiment procedure

1 The apparatus was set up as shown in Figure 5.7.
2 A wooden match was fixed to the end of each metal strip using a small amount of Vaseline.
3 The middle point of the ring was heated using the blue heating flame from a Bunsen burner.
4 The time taken for the match to fall off each metal strip was measured and recorded.

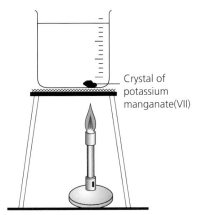

Figure 5.5 Convection experiment

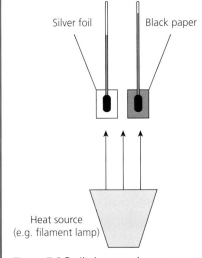

Figure 5.6 Radiation experiment

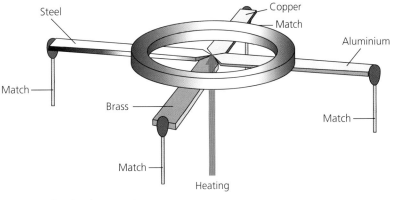

Figure 5.7 Conduction experiment

Convection experiment results

The student drew the observation shown in Figure 5.8.

Radiation experiment results

Thermometer	Temperature at the start of the experiment (°C)	Temperature after 10 minutes heating (°C)	Temperature difference (°C)
Black paper covered bulb	18	32	
Aluminium foil covered bulb	18	24	

Conduction experiment results

Metal strip	Time for wooden match to fall (s)
Copper	28
Steel	251
Brass	98
Aluminium	53

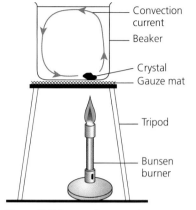

Figure 5.8 Convection experiment observation

Analysing your results

1 Explain the pattern of the movement of the purple dye from the potassium manganate(VII) crystal.
2 **a)** Calculate the temperature difference for each thermometer bulb.
   b) Which colour is the best absorber of radiated heat energy?
3 State the order of the heat conductivities of the metals, starting with the best conductor.

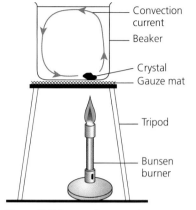

Figure 5.9 'Team Wales' coach

▶ Improving vehicle efficiency

Figure 5.9 shows the 'Team Wales' coach that is used to transport the national teams around the UK.

This version of the Wales team coach has a number of design features that are intended to increase the efficiency of the vehicle:

▶ The front top surface of the coach is angled and the edges of the outer surface are 'rounded'. This improves the **aerodynamics** of the coach, reducing the air resistance when the coach is travelling at speed.
▶ The bottom of the coach is close to the ground and the wheels are enclosed by the wheel arches. This also improves the aerodynamics of the coach.
▶ The hard material and high pressure of the tyres is designed to ensure a balance between the need for grip (to ensure the safety of the passengers) and the need to minimise the **rolling resistance** between the tyre and the road surface (improving the fuel economy).

The engine type affects the fuel economy of vehicles. Motors in electric cars are typically 85–90% efficient, whereas petrol engines are about 20% and diesel engines 30% efficient. Most coaches and buses still have diesel engines. Changing to an electric coach would substantially reduce the carbon footprint of Team Wales' transport.

Many newer vehicles improve their fuel economy by reducing the amount of energy lost when the vehicle is idling in traffic or at traffic lights. Computers monitor the engine and, when the vehicle stops, systems temporarily shut down the engine.

Similar systems are used in some petrol–electric hybrid cars to recover some of the kinetic energy of the engine that is wasted when the car is coasting (called inertia losses). Large complex vehicles, such as the Team Wales coach, use a variety of these measures to increase their fuel efficiency.

✔ Test yourself

4 What are the benefits from improving the aerodynamics of a coach by having an angled front top surface?
5 What are the two main ways to reduce the rolling resistance of a coach tyre?
6 What is the main benefit of shutting down the engine of a vehicle when it comes to a temporary stop?

Chapter summary

- Energy can be transferred by conduction, convection and radiation.
- The cost of using appliances can be investigated by using the efficiency energy banding system (A–G) and the power ratings of the appliances.
- Draught excluders, loft insulation, cavity wall insulation and double glazing can restrict energy losses from houses, reducing the carbon footprint and environmental impact of the home.
- The installation cost of the insulation and the fuel cost savings determines the cost effectiveness of introducing insulation systems into a house. The payback time is the duration (in years) before the savings start to outweigh the installation costs.
- The power of domestic appliances is given in kilowatts (kW). The kilowatt hour (kWh) is the unit of energy used by energy companies when charging customers.
- The cost of electricity can be calculated using the equations:

$$\text{units used (kWh)} = \text{power (kW)} \times \text{time (h)}$$
$$\text{cost} = \text{units used} \times \text{cost per unit}$$

- The energy efficiency of vehicles is mainly determined by the engine type but can be improved by: reducing aerodynamic losses (or air resistance); reducing the rolling resistance of tyres; and reducing idling losses.

Building electric circuits

All the electrical and electronic devices in our homes and businesses have electrical circuits. A basic understanding of how circuits work can help us use them efficiently and safely.

▶ Circuit symbols

A system of symbols and diagrams is used to represent components and connections in circuits. Standard circuit symbols are understood all over the world (Figure 6.1).

Cell/battery		Ammeter	(A)	A.C. power supply	—○ ～ ○—
Lamp	⊗	Voltmeter	(V)	D.C. power supply	+ −
Switch	—○ ⁄ ○—	Resistor	▭	Variable D.C. power supply	+ −
Variable resistor	▭	Fuse	▭	LDR	
Diode	⊕	Thermistor	⁄	LED	

Figure 6.1 Some common circuit symbols

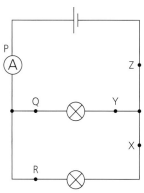

Figure 6.2 A series circuit

▶ Series and parallel circuits

The circuit symbols in Figure 6.2 show two lamps and an ammeter connected in series to a cell. (Note: a battery is two or more cells connected together.)

Figure 6.2 shows a **series circuit** and Figure 6.3 shows a **parallel circuit**. Both circuits have an ammeter connected in series with the battery to measure the **current**. Components are connected in series in a loop, so that the same current passes through each component. The current measured at A, B and C will be the same. In parallel circuits, two or more components are connected to the same points (called junctions) and the current splits, with some flowing through each component.

Current splits at a junction, so, in Figure 6.3, the current at P equals the current at Q plus the current at R.

The **current** in a circuit is a measure of the rate of flow of electricity around the circuit. Current is measured by connecting an ammeter into the circuit in series with other components. Current is measured in amperes (amps), with the symbol A.

Figure 6.3 A parallel circuit

Ⓗ

Test yourself

1 Study the circuits in Figure 6.4. Calculate the current at each of the marked points, **a** to **j**, on the circuit diagrams.

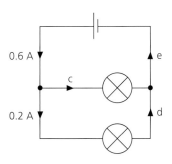
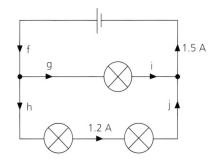

Figure 6.4

2 In a domestic house, all the electrical sockets and all the domestic appliances are connected in parallel to the main circuit board. The lights use 2.5 A, a television 0.5 A, an electric oven 13 A and a kettle 10 A. What is the total current drawn from the main circuit board?

► Voltage

Voltage is a measure of how much energy is being transferred by a component when it is working. Voltage is measured in volts, V, by a voltmeter. Voltmeters are connected in parallel with components, as shown in Figure 6.5.

In a series circuit, the voltages across each component add up to the supply voltage. In a parallel circuit, the voltage is the same across each branch of the circuit.

In Figure 6.2, the voltage across AB plus the voltage across BC equals the voltage across AC. In Figure 6.3, the voltage across PZ equals the voltage across QY, equals the voltage across RX.

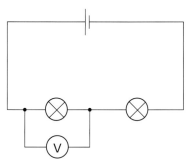

Figure 6.5 A voltmeter measuring voltage across one lamp in a circuit

Test yourself

3 Figure 6.6 shows a 12 V solar cell which is used to run three lamps, Bulb 1, Bulb 2, and Bulb 3.

a) Bethan connected a voltmeter across the solar cell. What voltage would she expect to measure during the day?

b) Explain why her voltmeter would read 0 V at midnight.

c) During the day, Bethan connected the voltmeter across points A and B in the circuit and turned on switches 1 and 2. What would her voltmeter read?

d) Explain why the lighting circuit has three switches. What does each switch do?

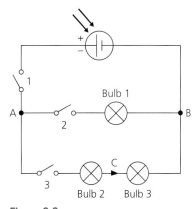

Figure 6.6

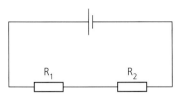

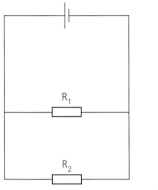

Figure 6.7 Two resistors in series

Figure 6.8 Two resistors in parallel

Resistance in circuits

The **resistance** of a component, measured in ohms, Ω, is its opposition to the flow of electricity flowing through it. Components in electronic circuits work at low current, so they tend to have high resistance. Insulating components, such as the casings of laptops or mobile phones, are made from very high resistance materials such as plastics.

Combining resistors in series and in parallel circuits

When two or more resistors are combined together in a series circuit, the total resistance of the circuit increases, and it is calculated by adding all the resistances. Figure 6.7 shows two resistors, R_1 and R_2, in series with a battery.

The total resistance of this circuit, R, is given by the equation:

$$R = R_1 + R_2$$

Combining resistors in parallel reduces the overall resistance of the circuit. Figure 6.8 shows two resistors, R_1 and R_2, arranged in parallel with a battery.

In parallel circuits, the combined resistance of two resistors is always less than the resistance of the lowest resistor.

The overall resistance, R, of a parallel circuit can be calculated using the equation:

$$\frac{1}{R} = \frac{1}{R_1} + \frac{1}{R_2}$$

★ **Worked example**

Calculate the total resistance of a series circuit containing a 32 Ω resistor and a 14 Ω resistor.

Answer

The total resistance $R = R_1 + R_2 = 32\,\Omega + 14\,\Omega = 46\,\Omega$

★ **Worked example**

A parallel circuit consists of an 8 Ω resistor in parallel with a 4 Ω resistor. Calculate the overall resistance of the circuit.

Answer

$$\frac{1}{R} = \frac{1}{R_1} + \frac{1}{R_2} = \frac{1}{8\,\Omega} + \frac{1}{4\,\Omega}$$

$$= 0.375/\Omega \implies R = \frac{1}{0.375} = 2.67\,\Omega$$

✓ **Test yourself**

4 Calculate the total resistance of a series circuit containing a 4.6 Ω resistor, a 1.3 Ω and a 2.6 Ω resistor.

5 Two 6 Ω resistors are connected in parallel. Calculate the overall resistance of the circuit.

Modern living and energy

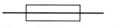

Figure 6.9 A standard 13 A fuse and its circuit symbol

▶ Fuses

The plugs of appliances are fitted with a safety device called a **fuse** (Figure 6.9). If the current drawn from the mains socket exceeds the maximum current rating of the fuse, the wire within the fuse melts, disconnecting the mains supply. Fuses are fitted to the live wire of the plug, preventing too much current from flowing and causing a fire.

Melted fuses must be replaced before the appliance can be used again. The rating of a fuse (in amps) is always higher than the normal operating current of the mains appliance:

- ▸ 3 A fuses (red) – for appliances up to 700 W
- ▸ 5 A fuses (black) – for appliances from 700 to 1200 W
- ▸ 13 A fuses (brown) – for appliances from 1200 to 3000 W.

✔ Test yourself

6 Explain how a fuse works.
7 A hairdryer has a stated electrical power of 1300 W. Calculate the rating of the fuse that should be fitted to its plug. Explain your choice.

★ Worked example

Calculate the current drawn from a 2.5 kW mains kettle, operating with a voltage of 230 V, and use this value to decide on the rating of the fuse for the plug. Use the equation:

$$\text{current} = \frac{\text{power}}{\text{voltage}}$$

Answer

$$2.5\,\text{kW} = 2500\,\text{W}$$

$$\text{current} = \frac{\text{power}}{\text{voltage}} = \frac{2500\ \text{W}}{230\ \text{V}} = 10.9\ \text{A}$$

A 5 A fuse is too low, but a 13 A fuse would be suitable.

▶ Ohm's law

In 1827 Georg Ohm discovered that the size of the current passing through a wire was directly proportional to the voltage applied across it – if the voltage across it is doubled, then the current through it will also double. Ohm also varied the dimensions and material of the wires and found that if he kept the voltage constant, the current was inversely proportional to the resistance of the wire, meaning that if he doubled the resistance of the wire (by doubling its length), then the current halved. Today we summarise Ohm's findings using the equation:

$$\text{current} = \frac{\text{voltage}}{\text{resistance}} \quad \text{or} \quad I = \frac{V}{R}$$

This relationship can be written in several other ways:

$$V = IR \quad \text{and} \quad R = \frac{V}{I}$$

✔ | **Test yourself**

8 A 25 Ω fixed resistor has a current of 2 A through it. Calculate the voltage across the fixed resistor.

9 Figure 6.10 shows the electrical characteristic of a 12 V car lamp. Use the graph to calculate the resistance of the lamp when the current through the lamp is:
   a) 0.2 A
   b) 0.6 A
   c) 1.0 A.

Ⓗ 10 A rheostat (large variable resistor) is set up with a resistance of 12 Ω. A 0–12 V variable power supply is connected in series with an ammeter and the rheostat. Determine the current through the rheostat, for voltages of 0 V, 4 V, 8 V, and 12 V.

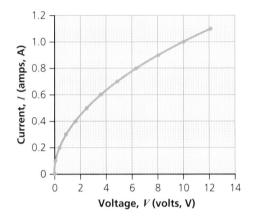

Figure 6.10 The *I–V* graph for a car lamp

► The electrical characteristics of components

Key term

Electrical-characteristic graph
   Current–voltage graph.

The electrical-characteristic (*I–V*) graphs of a fixed resistor and a filament lamp are shown in Figure 6.11 (a) and (b).

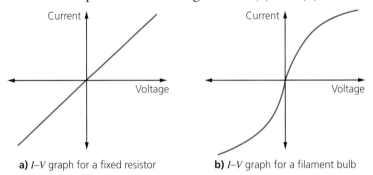

a) *I–V* graph for a fixed resistor   b) *I–V* graph for a filament bulb

Figure 6.11 Electrical-characteristic graphs (*I–V*) of a) a fixed resistor and b) a filament bulb

Fixed resistors obey Ohm's law and their *I–V* graphs are linear (straight lines). The higher the resistance of the resistor, the shallower the slope of the *I–V* graph. Filament lamps heat up as more current passes through them. Increasing the temperature of the filament causes the resistance of the filament to increase. An *I–V* electrical-characteristic graph for a filament lamp is a curve with a decreasing slope, as shown in Figure 6.11 (b).

Notice that electrical characteristics are usually drawn with positive and negative voltages, this is because some components (like diodes) behave differently depending on the direction in which they are connected.

Modern living and energy

Investigation of the current-voltage (I–V) characteristics for a component

A student performed an experiment to investigate how the current varied with voltage for a 12 V filament lamp. A circuit diagram of the apparatus that the student used is shown in Figure 6.12.

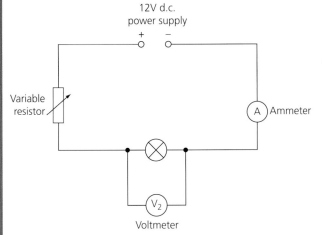

Figure 6.12 Circuit diagram for an experiment to determine the I–V characteristic for a 12 V filament bulb

Procedure

1 The apparatus was connected up as shown in the circuit diagram, Figure 6.12.
2 The variable resistor was altered until the reading on the voltmeter measured 1 V.
3 The values of current and voltage were measured and recorded.
4 Steps 2 and 3 were repeated for voltages up to 12 V in 1 V intervals.

Results

voltage, (V)	0.0	1.0	2.0	3.0	4.0	5.0	6.0	7.0	8.0	9.0	10.0	11.0	12.0
current, (I)	0.0	1.3	2.4	3.4	4.1	4.6	5.0	5.3	5.5	5.7	5.8	5.9	6.0

Analysing your results

1 Plot a graph of current (on the y-axis) against voltage (on the x-axis).
2 Draw a line of best fit on your graph.
3 The student suggested that if she set the power supply to 3.5 V, then the current would be 4.0 A. State, with a reason, if you think that the student is correct.
4 The student's teacher suggested that she turns the connections to the power supply around and repeats the experiment with negative voltages. Draw a sketch of the complete graph including both positive and negative voltages.

Figure 6.13 The circuit symbol for a diode

Diodes

Diodes are electrical components that allow electricity to pass through them in one direction only. The electrical symbol for a diode is shown in Figure 6.13.

Diodes are used in a.c. to d.c. converters and as room, external or vehicle lighting. The electrical characteristic of a diode is shown in Figure 6.14.

The I-V graph for a diode shows that for positive voltages, the diode conducts electricity. If the power supply is reversed, and negative voltages applied, then the diode does not conduct at all.

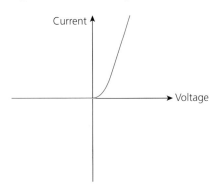

Figure 6.14 The electrical characteristic of a diode

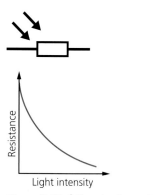

Figure 6.15 a) The circuit symbol for a LDR, **b)** the variation of the resistance of a typical LDR with light intensity

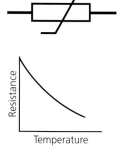

Figure 6.16 a) The circuit symbol for a thermistor, and **b)** the variation of resistance with temperature for a typical thermistor

Light-dependent resistors and thermistors

Light-dependent resistors (LDRs) are components that change resistance with the intensity of the light. Many LDRs have decreasing resistance with increasing light intensity. Figure 6.15 shows the variation of the resistance of a typical LDR with light intensity.

Thermistors are electrical components that change their resistance with temperature. They are frequently used as electrical temperature sensors. Most thermistors decrease their resistance as their temperature increases, as shown in Figure 6.16.

▶ Electrical power

Electrical power is the rate at which a device transfers electricity into other forms; in other words, how much energy the device can transform per second:

$$\text{power} = \frac{\text{energy transfer}}{\text{time}}$$

Very powerful electrical devices, such as mowers, can convert a large amount of electrical energy per second into other useful forms. Electrical power is measured in watts, W.

The electrical power of a device can also be calculated by multiplying the voltage and the current of the device together:

$$\text{electrical power} = \text{voltage} \times \text{current}$$

or

$$P = VI$$

Current, resistance and power

(H) If $P = VI$ and $V = IR$ (from Ohm's law), then substituting for V in the power equation gives:

$$P = (IR) \times I = I^2R$$

or

$$\text{power} = \text{current}^2 \times \text{resistance}$$

This is a useful equation for calculating the power consumption in complex circuits. Measure the current flowing through a component, square this value, then multiply by its resistance, and you have the power of the component.

★ Worked examples

1 Calculate the power of a 12 V light bulb with a current of 0.5 A flowing through it.

 2 A 230 V mower has a power of 2000 W. Calculate the current flowing through the mower.

3 A 400 Ω resistor has a current of 0.75 A flowing through it. Calculate the power of the resistor.

Answers

1 $P = VI = 12\,\text{V} \times 0.5\,\text{A} = 6\,\text{W}$

 2 $P = VI$ so $I = \dfrac{P}{V} = \dfrac{2000\,\text{W}}{230\,\text{V}} = 8.7\,\text{A}$

3 $P = I^2R = (0.75)^2\,\text{A} \times 400\,\Omega = 225\,\text{W}$

✔ Test yourself

11 Calculate the power of a 6 V torch lamp drawing a current of 0.8 A.

⬇ Chapter summary

- Electrical circuit symbols are used to represent circuits on circuit diagrams.
- In series circuits, the components are connected in a loop so that the current is the same through each component, and the supply voltage is the sum of the voltages across each component.
- In parallel circuits, there are branches, and two or more components are connected to the same points in the circuit. The sum of the currents in each branch is equal to the current from the supply, and across each branch, the voltage is the same.
- Current is measured by an ammeter, connected in series in a circuit.
- The ampere (or amp), A, is the unit for current.
- Voltage is measured in volts, V, by a voltmeter connected in parallel across a component.
- Resistance is measured in ohms, Ω.
- Adding components in series increases the total resistance of a circuit; adding components in parallel decreases the total resistance of a circuit.
- The resistance of a thermistor decreases with increasing temperature. The resistance of a LDR decreases with increasing light intensity.

- The total resistance, R, of two resistors connected in series is calculated by: $R = R_1 + R_2$
- The total resistance, R, of two resistors connected in parallel can be calculated by the equation:

$$\frac{1}{R} = \frac{1}{R_1} + \frac{1}{R_2}$$

- The current flowing through a component depends on the voltage applied across it: the larger the voltage, the larger the current.

$$\text{current} = \frac{\text{voltage}}{\text{resistance}} \quad \text{or} \quad I = \frac{V}{R}$$

- The current–voltage (I–V) characteristic graphs for a wire, filament lamp and a diode are distinctive shapes and can be used to identify the component.
- Electrical power, P, is measured in watts, W, and $P = VI$ or $P = I^2R$
- The electrical fuse for a device's plug needs to be chosen so that it is just bigger than the normal operating current for the device.

► Practice exam questions

1 The graph shows how the combined electrical power usefully transferred by all the wind farms in North Wales changed on 2nd July.

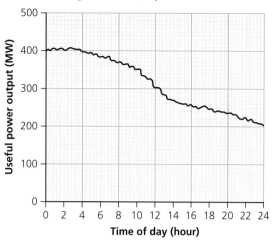

a) Describe how the strength of the wind changed throughout the day. [1]

b) i) Use the graph to determine the combined electrical power usefully transferred at 12.00 hours. [1]

ii) The total power supplied at 12.00 hours to the wind farms was 600 MW.

Use the equation:

$$\%\text{efficiency} = \frac{\text{power usefully transferred}}{\text{total power supplied}} \times 100$$

to calculate the combined efficiency of all the wind farms in North Wales. [2]

2

Not to scale

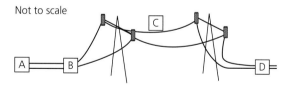

a) A section of the National Grid is shown. Box A represents a solar power farm.

Use a word from the box below to copy and complete the sentences that follow. Each word may be used once, more than once or not at all.

power	generator	pylon	transformer	current

i) At B, the voltage is increased using a ... . [1]

ii) Energy losses at C are minimised because the ... is reduced. [1]

iii) At D, a ... is used to decrease the voltage. [1]

b) The demand for electricity on Anglesey during one day in June, is shown.

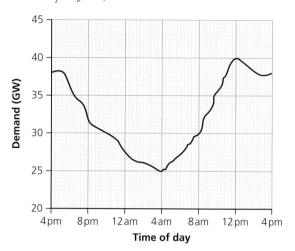

i) Use the graph to determine the maximum demand. [1]

ii) Between 4 am and 12 pm, the demand for electricity increased. Use the graph to determine this increase. [1]

c) A schematic diagram of Dinorwig pumped hydroelectric power station is shown. At night, water is pumped from the lower to the upper reservoir, using electricity from the National Grid.

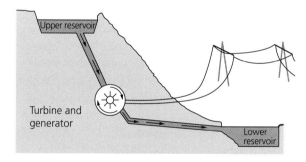

i) Suggest a reason why the water is pumped up to the top reservoir at night. [1]

ii) Dinorwig was used to supply Anglesey with electricity between 11 am and 12 pm. Suggest a reason why. [1]

3 The circuit diagram may be used to find the resistance of a fuse used as part of a model electric train set.

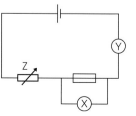

a) Identify components X, Y and Z. [3]

b) Use the equation:

$$\text{resistance} = \frac{\text{voltage}}{\text{current}}$$

to calculate the resistance of the fuse if its voltage is 6 V and the current is 1.5 A. [2]

c) The fuse is put into the train track lighting circuit shown below.

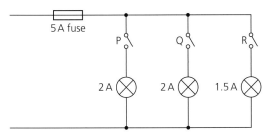

i) Calculate the total current drawn from the supply when switches P, Q and R are pressed ON. [2]

ii) State what will happen to the fuse and the lamps when these three switches are pressed ON. [2]

4 A homeowner has a maximum of £3000 available to spend on improving home insulation. Information on each type of insulation is shown in the table.

Part of house	Insulated or not	Energy lost per second (W)	Cost of insulation (£)	Payback time (years)	Expected annual saving (£ per year)
Loft	None	4500			
	Thick fibre glass blanket	1200	600	X	200
Cavity wall	None	3400			
	Expanded foam	1400	1800	10	180
Doors	Wood	1000			
	uPVC plastic	800	1500	50	Y
Windows	Single glazing	1800			
	Double glazing	1400	2400	60	40

a) Use the equations below:

$$\text{payback time (years)} = \frac{\text{cost of insulation (£)}}{\text{annual savings (£ per year)}}$$

$$\text{annual savings (£ per year)} = \frac{\text{cost of insulation (£)}}{\text{payback time (years)}}$$

to calculate the missing values X and Y. [2]

b) Use information from the table to advise the homeowner on how best to spend a maximum of £3000 on insulation. [6]

c) Use the table to calculate the energy loss per second from the house when it is:

i) Un-insulated [1]

ii) Fully insulated [1]

iii) Use your values from c) i) and ii) to determine the percentage decrease in energy loss from the un-insulated house to the fully insulated house. [2]

5 David investigated how the current through a 12 V filament lamp varied with the voltage across it, from 0 and 12 V.

a) Part of the circuit that David used is shown in the diagram. Copy the diagram and add a voltmeter and an ammeter to the circuit. [2]

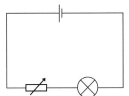

Variable resistor

b) Describe the method that the student could use to take a range of results. [3]

c) The student obtained the following readings of current against voltage for the lamp.

Voltage (V)	Current (A)
0.0	0.0
0.5	0.8
2.0	1.6
4.0	2.1
6.0	2.4
8.0	2.6
10.0	2.8
12.0	2.9

i) Plot the data on a grid and draw a suitable line of best fit. [3]

ii) Describe the pattern in the data shown on the graph. [2]

d) The student is told that the standard operating voltage of the lamp is 12 V. Use the equation:

$$\text{power} = \text{voltage} \times \text{current}$$

to calculate the power of the lamp at 12 V. [2]

e) A similar lamp has a resistance of 4.8 Ω when the current flowing through it is 3.0 A. Use the equation:

$$\text{power} = \text{current}^2 \times \text{resistance}$$

to calculate the power of this lamp at a current of 3.0 A. [2]

f) The student repeated this experiment with a fixed resistor. The current was 1.5 A when the voltage across it was 8 V. Add a line to your graph to show how current varies with voltage for the fixed resistor. [2]

7 Obtaining clean water

The compound water contains atoms of the elements oxygen and hydrogen chemically bonded together. It is essential for human life as our bodies are about 60% water. Although we can live for weeks or months without food, we can only survive about three days without water. Water must be safe to drink, because it can carry pathogens that cause disease and dissolved substances that are poisonous.

▶ Atoms and elements

All substances are made from atoms

Atoms are the smallest particle that can exist on its own. Most of the mass of the atom is from the protons and neutrons which are found in the central nucleus of the atom.

All atoms of the same element have the same number of protons in their nucleus and so have the same atomic number. Hydrogen is an element with one proton in the nucleus and so has an atomic number of 1. Oxygen has 8 protons in the nucleus and an atomic number of 8.

The Periodic Table lists all known elements and includes the atomic number and mass number of the atoms of each element. Every element has its own symbol of one or two letters.

★ **Worked example**

Use the information in Figure 7.1 to determine the number of each subatomic particle in a sodium atom.

Mass number —— $^{23}_{11}$**Na**
Atomic number ——

Figure 7.1 The full symbol for a sodium atom which contains information to determine the number of each subatomic particle

Answer

The atomic number is 11, telling us that there are 11 protons in the nucleus.

A neutral atom has the same number of electrons as protons. So, there must also be 11 electrons.

The mass number is 23, as only protons and neutrons have significant mass, so we can work out the number of neutrons by taking away the atomic number from the mass number:

$23 - 11 = 12$ neutrons in the nucleus.

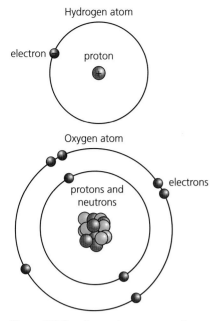

Hydrogen atom

electron — proton

Oxygen atom

protons and neutrons — electrons

Figure 7.2 The electronic structure of hydrogen is 1 and the electronic structure of oxygen is 2,6

Electronic structure

Electrons orbit the nucleus in shells. The electrons are filled from the inner shell outwards. The inner shell can hold up to two electrons, then the next two shells can hold a maximum of eight electrons. Figure 7.2 shows the electronic structure of hydrogen and oxygen.

✔ **Test yourself**

1 Lithium has the symbol: $^{7}_{3}\text{Li}$.
   a) Calculate the number of each subatomic particle in an atom of lithium-7.
   b) Give the electronic structure of lithium.
2 Hydrogen has the symbol: $^{1}_{1}\text{H}$.
   a) Name the subatomic particle that is not present in this atom of hydrogen.
   b) Describe the position of the electron in this atom.
3 Chlorine has the symbol: $^{35}_{17}\text{Cl}$.
   a) Determine the mass number of this atom of chlorine.
   b) Describe the position of the neutrons in this atom.

Isotopes

Key term

Isotope Atoms of the same element with the same atomic number but different mass number.

Isotopes are different forms of the same element, so they have the same atomic number. As they have different numbers of neutrons in the nucleus, they have different mass numbers. Figure 7.3 shows isotopes of hydrogen. All of the atoms have the same number of protons, so they are the same element. However, they have different numbers of neutrons and so they also have different mass numbers. The chemical reactions of isotopes are the same as they have the same number and arrangement of electrons, but the physical properties, such as melting point and density, are different due to the different masses.

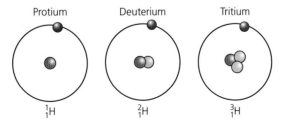

Protium   Deuterium   Tritium

$^{1}_{1}\text{H}$ $^{2}_{1}\text{H}$ $^{3}_{1}\text{H}$

Figure 7.3 The atomic structure of all the isotopes of hydrogen

✔ **Test yourself**

4 Chlorine has two isotopes: $^{35}_{17}\text{Cl}$ and $^{37}_{17}\text{Cl}$. Describe how these isotopes are similar and how they are different.
5 Explain why isotopes of the same element have the same chemical properties.
6 Explain why isotopes have slightly different physical properties.

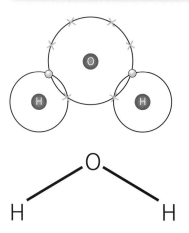

Figure 7.4 Dot and cross diagram and structural formula to show the bonding in a water molecule

► Compounds

Elements are the building blocks of all substances. They are substances that cannot be broken down by chemical means but they can join or bond to other elements to make **compounds**.

Covalent compounds

Water molecules are made from two hydrogen atoms bonded to one oxygen atom. Figure 7.4 shows two different representations of a molecule of water:

► a dot and cross diagram with oxygen's outer shell electrons as dots and the single electron of each hydrogen atom as a cross. The atoms are bonded together by a shared pair of electrons which form a **covalent bond**.
► a structural formula showing the bond between the atoms.

You will learn more about covalent bonding in Chapter 18 Materials for a purpose.

We can use symbols to write a molecular formula for water: H_2O, where the 2 shows there are two atoms of hydrogen, H, for every one atom of oxygen, O.

✔ Test yourself

7 Oxygen and hydrogen can also make molecules of hydrogen peroxide. Give the molecular formula for the compound made from two hydrogen atoms and two oxygen atoms chemically joined.
8 Draw the dot and cross diagram for hydrogen peroxide.
9 Draw the stick diagram for hydrogen peroxide.

Ionic compounds

Atoms can lose or gain electrons to be become **ions**.

When oxygen forms compounds with metals it makes an ionic bond. For example, magnesium is in group 2 and will lose its outer two electrons to make a 2+ ion with the formula Mg^{2+}. This is formed by the electrostatic force of attraction between oppositely charged ions. As oxygen is a non-metal it always forms the negative ion and is attracted to positive metal ions to form the ionic compound. Oxygen always gains two electrons to become an oxide ion with a −2 charge. The symbol for an oxide ion is O^{2-}. You will learn more about ionic bonding in Chapter 18 Materials for a purpose.

✔ Test yourself

10 Give the symbol for the following ions:
   a) A calcium atom that has lost two electrons
   b) A magnesium atom that has lost two electrons
   c) A chlorine atom that has gained one electron

▶ Pure water and 'natural' water

Pure water contains only water molecules. 'Natural' water is not pure as it includes dissolved gases, ions including metal ions, carbonates and nitrates, particulate matter, pathogens organic matter and pesticides.

Natural water supplies include:

- ▶ salt water in seas and oceans
- ▶ fresh water in streams and rivers
- ▶ aquifers (ground water in permeable rock) accessed by bore holes or wells.

Distillation, desalination and reverse osmosis

Natural water can be purified by distillation. Figure 7.5 shows a distillation set-up.

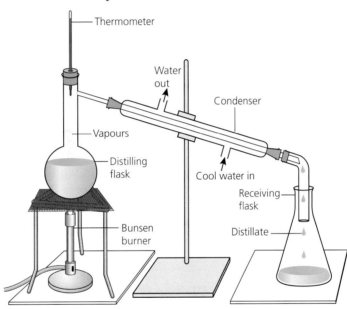

Figure 7.5 Experimental set-up for a simple distillation to extract the solvent from a solution

In Figure 7.5 sea water is put into the round bottom flask. It is heated so that the water evaporates and all the dissolved substances, including any miscible liquids, are left behind. The thermometer takes the temperature of the vapour, which is also the boiling point of the steam (100 °C). This exact boiling point indicates that the vapour is pure water. The steam travels into the condenser and cools into pure liquid water, which is collected in the conical flask.

Desalination can be used to produce pure drinking water from sea water. Drinking water is described as **potable** water. Desalination by distillation uses such a lot of energy that it is unsustainable on a large scale without access to large quantities of renewable energy.

Reverse osmosis is another method to desalinate and purify water. The water is passed through a selectively permeable membrane under high pressure. Only the water molecules fit through the pores in the membrane, leaving most of the other ions and molecules that were dissolved in the water behind. This method uses expensive membranes and is inefficient because a lot of waste water is produced. It is used on many ocean-going ships, such as the HMS Queen Elizabeth, the flagship of the UK's Royal Navy, to produce potable water.

▶ Sustainable water supplies

Water is a **finite resource**. This means we need to use water sustainably to ensure that we can live our lives as we want to now, while also preserving supplies for future generations. To use water resources sustainably we can:

- reduce domestic water consumption by buying appliances that use less water, thinking about the frequency of laundry and choosing drought tolerant plants in gardens.
- reduce commercial water consumption by growing crops that need less water and, in industry, recycling water in factories.
- treat waste water to make it potable, rather than using natural water supplies.

By using water sustainably we can reduce the impact of water storage, extraction and treatment on the environment.

Extracting and storing water for human use from natural sources can change ecosystems. For example, the Elan Valley Reservoirs in Mid Wales are used as storage for drinking water that is supplied to England. The reservoirs were made in the 1800s by damming the Elan and Claerwen rivers. This flooded the habitats on either side of the river and has also changed the amount of water flowing into the sea. Aqueducts were built to transport the water from Wales to England.

Wells are holes in the Earth that fill with ground water. Bore holes are deep wells, which are dug down to meet the ground water. The bore hole at Morfa Bychan in Gwynedd, North Wales, is used as a source for drinking water for the customers of Welsh Water. Lowering of the water table through use of wells and bore holes affects ecosystems as there is less available water, and they can also lead to subsidence.

▶ Water treatment

Natural water supplies contain dissolved substances from the rocks that the water passed over. These are usually present at safe levels. However, pollution can affect natural sources of water:

- Fertilisers can run off the fields and add extra nitrates.
- Pesticides can enter the water, adding dissolved chemicals that would not normally be there.
- Factories can release dissolved metal ions.

Dangerous levels of dissolved substances can lead to illness. It is important that water is tested and the levels of potentially harmful substances monitored. Industries are also monitored to prevent the release of the substances into water sources such as rivers and oceans.

In the UK, drinking water is produced in water treatment plants and pumped to every home. Figure 7.6 shows a flow chart of how the drinking water is made.

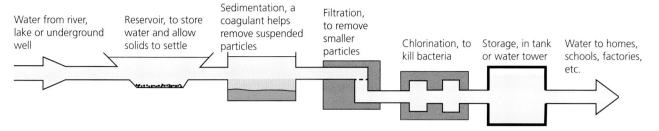

Figure 7.6 Stages in the treatment of drinking water

Raw natural water is collected and pumped into the water works. Then coagulants are added to make the particulate matter stick together into flocks that sink to the bottom of the sedimentation tank and are removed. A polymer is added to help remove any smaller insoluble particles by filtration. Fluoride is added to some water supplies to improve the dental health of the population. The final step is to ensure the water is safe to drink by adding disinfectant gases such as chlorine or ozone. These kill pathogens. The water is stored in vast underground tanks until it is needed in homes, schools and workplaces.

> **Test yourself**
>
> 14 Give the main three stages of water treatment to make drinking water.
> 15 List two disinfection agents that can be added to drinking water.
> 16 Give the formula of a fluoride ion.

▶ Hard and soft water

Soft water has no dissolved salts in it. Rainwater is soft water. As rainwater flows over rocks, some of the mineral salts dissolve. **Hard water** contains dissolved magnesium and calcium ions which stop soap foaming. So, whether water is hard or soft is determined by the geology of the region. Homes in Snowdonia have soft water as their water supply is from rainfall into upland reservoirs, whereas Llandrindod Wells has hard water from underground springs.

Water hardness can be:

> **Key term**
>
> **Hard water** Water with dissolved calcium or magnesium salts that does not foam with soap.

▶ temporary – caused by dissolved calcium hydrogen carbonate ($Ca(HCO_3)_2$).
▶ permanent – caused by dissolved calcium sulfate ($CaSO_4$).

Table 7.1 Comparison of the characteristics of soft and hard water

Soft water	Hard water
• Does not contain calcium or magnesium ions • Has no taste • Foams easily with soap • Foams with a detergent	• Contains calcium and/or magnesium ions • Many people prefer the taste • Does not foam easily with soap • Foams with a detergent • Health benefits as contains calcium ions which are good for bones and heart • Causes limescale which can build up and lead to burst pipes and reduced efficiency of heating elements in kettles and boilers

Table 7.2 Mechanisms, advantages and disadvantages of different methods of water softening

Method	Effect	Disadvantages	Advantages
Boiling	High temperature causes the calcium hydrogen carbonate to undergo thermal decomposition into calcium carbonate which is insoluble and can be filtered out	Only suitable for temporary hardness. Can cause limescale build up on heating elements, reducing their effectiveness and shortening lifespan.	Simple procedure. No additives. Potable water produced.
Adding washing soda	Calcium carbonate precipitate forms and is washed away	Only suitable for cleaning appliances, such as washing machines and dishwashers, as sodium ions are added. Irritant chemical. Sodium ions enter waste water.	Inexpensive additive. Simple to use. Better cleaning performance, so less detergent required. Household appliances last longer. Less energy is needed to heat water if there is no limescale on the heating element. Can remove both temporary and permanent hardness.
Ion exchange	Calcium and magnesium ions are swapped for hydrogen or sodium ions	Requires specialised equipment that has to be maintained. Sodium ions in drinking water are linked to health concerns such as high blood pressure. Sodium ions enter waste water.	Household appliances last longer. Household pipes not affected by limescale build up. Can soften large amounts of water. Better cleaning performance, so less detergent required. Less energy is needed to heat water if there is no limescale on heating elements. Can remove both temporary and permanent hardness. Potable water produced.

Key terms

Thermal decomposition The breaking down of a substance into simpler substances using heat.

Precipitate An insoluble solid produced during a chemical reaction in solution.

✔ Test yourself

17 Give one benefit of hard water and one benefit of soft water.

18 Explain how living in a hard water area can be more expensive than living in a soft water area.

⚙ Specified practical

Determination of the amount of hardness in water using soap solution

Water companies regularly take samples of water and complete a number of tests to describe the water purity and quality. Measuring the hardness of water is important so that people and businesses can consider if it is worth the cost to soften the water or not.

In this experiment, a student was asked to determine the hardness of a water sample using a soap solution method.

Procedure

1 Water sample A was measured into a conical flask.
2 1 cm³ of soap solution was added, the stopper was inserted and the flask was shaken vigorously for 5 s.
3 Step 2 was repeated until a lather formed and lasted for 30 s. The total volume of soap solution added was recorded.
4 Steps 1–3 were repeated with 50 cm³ samples of all other types of water.

The student concluded that the water samples that produce a foam after one drop of soap are soft water. The more drops of soap that are needed to make a foam, the harder the water.

Questions

1 In this investigation, state:
   a) the independent variable
   b) the dependent variable
   c) three control variables.
2 What safety precautions, if any, do you need for this experiment and why?
3 How could you improve the resolution (the change in the quantity) of the results?
4 How could you make the results more reliable (similar each time they are repeated)?

▶ Solubility

A solution is a colourless liquid made when a solute like sodium chloride dissolves into a solvent like water. Different amounts of substances can dissolve in the same volume of water. Solubility is the mass of a solute that can dissolve in a volume of solvent. It depends on temperature and the characteristics of the solute. In general:

▶ Solubility of solids increases with temperature.
▶ Solubility of gases decreases with temperature.

When no more solute can dissolve we describe the solution as saturated. You know a solution is saturated when you can see some of the solid at the bottom of the solution.

→ Activity

Solubility curves

It is possible to find the solubility of a solid like sodium carbonate by measuring the mass that can dissolve in a set volume of water at different temperatures.

The results of an investigation are shown in the table below.

Temperature (°C)	Mass (g)
0	0.7
10	1.3
20	2.2
30	4.0
40	4.9

Questions

1 Use the data to plot a solubility curve. Remember that the independent variable (temperature) should be on the x-axis and the dependent variable (mass) should be on the y-axis. Make sure you add a line of best fit, which should show the trend in the data.
2 Why do you need at least five results to draw a line of best fit?
3 Why can't you measure the solubility of sodium carbonate at −10 °C?
4 Why can't you measure the solubility of sodium carbonate at 110 °C?
5 What mass of sodium carbonate dissolved at 25 °C?

▼ Chapter summary

- Elements are made up of one type of atom and cannot be broken down by chemical means.
- Isotopes of the same element have the same atomic number (same number of protons) but a different mass number (different number of neutrons).
- Atoms can be described in terms of their atomic number, which tells you the number of protons in the nucleus and identifies the element. The mass number gives information about the number of protons and neutrons in the nucleus of the atom.
- Electrons occupy shells around the nucleus. The maximum number of electrons in the first shell is two, then eight in the next two shells.
- Ions are charged atoms or groups of atoms.
- Compounds are made of more than one type of atom that are chemically joined.
- Natural water is not pure as it contains dissolved gases, ions including metal ions, carbonates and nitrates, particulate matter, pathogens, organic matter and pesticides.

- Water is a finite resource and we must use it sustainably.
- Drinking water is made in treatment plants using natural water and treating it by using sedimentation and filtration to remove insoluble substances and disinfecting it with ozone or chlorine.
- Drinking water can be made by desalination of sea water using distillation or reverse osmosis.
- Pure water can be made by distillation.
- Different masses of substances dissolve at different temperatures in water. Solubility curves can be created to show how solubility changes with temperature.
- Dissolved magnesium and calcium ions can make water hard.
- Hard water can be softened by boiling, adding sodium carbonate or ion exchange.
- Hard water has a taste and health benefits. However, it can lead to limescale which can cause burst pipes and reduce the energy efficiency of heating elements.

Our planet

Raw materials

Our planet Earth contains the resources we need. We use naturally occurring **raw materials** like metals, rocks and minerals, processing them into useful materials. When we extract and process raw materials, we need to weigh up their benefits against any negative impacts on the environment.

The Periodic Table

All matter is made from **elements** that are chemically combined or mixed together. The elements are arranged in the Periodic Table so we can make predictions about their physical and chemical properties. They are classified as metals or non-metals based on their physical properties. Most metal elements are found to the left and centre of the Periodic Table. The non-metals are found to the right (Figure 8.1).

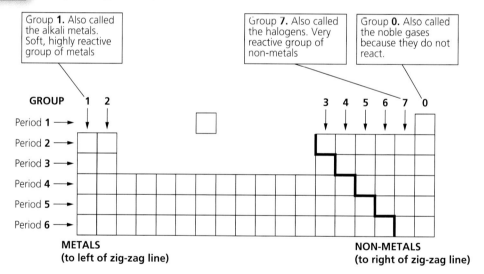

Group **1**. Also called the alkali metals. Soft, highly reactive group of metals

Group **7**. Also called the halogens. Very reactive group of non-metals

Group **0**. Also called the noble gases because they do not react.

METALS (to left of zig-zag line)

NON-METALS (to right of zig-zag line)

Figure 8.1 The Periodic Table

Rows in the Periodic Table are called **periods** and columns are called **groups**. All elements in the same group have similar chemical properties as they have the same electronic arrangement in their outer shell.

> **✓ Test yourself**
>
> 1 What is an element?
> 2 Where are the metal elements found on the Periodic Table?
> 3 Where are the non-metal elements found on the Periodic Table?
> 4 What is the difference between a group and a period of elements?

Group 1

Group 1 elements, also known as the alkali metals, are so reactive they are kept under oil or as an inert gas to stop them reacting. They are only found in compounds in the Earth's crust, but can be purified and extracted using physical and chemical processes. Like all metals, they are conductors, lustrous (shiny), ductile (can be drawn into wires) and malleable (bend easily).

Figure 8.2 Group 1 metals, such as lithium, sodium and potassium

→ Activity

Group 1

Table 8.1 Group 1 metals

Element	Melting point (°C)	Density (g/cm³)	Reactivity with water
Li	181	0.534	Floats, fizzes
Na	98	0.971	Floats, melts into a ball, fizzes
K	64	0.862	Floats, ignites with a purple flame, fizzes
Rb	39	1.532	White sparks and explodes on contact with water
Cs	28	1.873	Explodes

1 State the trend in melting point.
2 Use your knowledge of structure and bonding to explain the trend in melting point.
3 Suggest the anomalous value for the density data for Group 1 metals.
4 Justify which Group 1 metals float in water and which sink in water. (Density of water is about 1 g/cm³)

Group 7

The Group 7 elements are also known as the halogens. These non-metal elements are diatomic molecules.

All halogens are reactive. Fluorine is the most reactive non-metal in the Periodic Table. Reactivity decreases as you go down the group. The halogens are so reactive that their pure forms cannot be found in nature. They must be extracted using physical and chemical means. The Welsh Bromine Works based in Anglesey extracted bromine from sea water for use as a petrol additive in cars.

The colour of the first three halogens darkens as you go down the group: fluorine is a pale-yellow gas, chlorine is a green gas and bromine is an orange liquid.

Figure 8.3 The first three halogens

Group 7

The table shows data for the Group 7 halogens.

Element	Atomic radius (pm)	Melting point (°C)	Boiling point (°C)
F	71	−220	−188
Cl	99	−102	−34
Br	114	−7	59
I	133	113	184

1 Give the state of each element at room temperature (25 °C).
2 Astatine is below iodine in Group 7. Predict the melting point, boiling point and colour of this element.
3 Explain why these elements are all in the same group of the Periodic Table.
4 The unit, pm, is a picometre $(1 \times 10^{-12}\,m)$. Give the atomic radius of iodine in metres using standard form to two significant figures.

Test yourself

5 Which groups in the Periodic Table are the alkali metals and halogens found in?
6 List the properties that alkali metals have in common with all metals.
7 Describe the trend in reactivity of the Group 1 elements.
8 Describe the trend in melting point of the Group 7 elements.

Chemical reactions

In a chemical reaction a new substance is made; no atoms are created or destroyed, they are just rearranged. This means that mass is conserved (stays the same).

We can model what happens in a chemical reaction:

▶ The bonds in the starting chemicals or reactants break apart.
▶ The atoms rearrange into a new order to make the products.
▶ New bonds are formed to make the products.

Equations

Chemical changes are summarised in equations. The starting chemicals (reactants) are written on the left and the products are written on the right. We use → (never =) to show that reactants become products.

You should be able to write and interpret:

1 Word equations containing the names of all the substances involved in the reaction. For example:

magnesium + oxygen → magnesium oxide

sodium + chlorine → sodium chloride

2 Balanced symbol equations containing numbers in front of the chemical formulae to track the number and type of atoms involved in the reaction. For example:

$$Mg + O_2 \rightarrow 2MgO$$

$$Na + Cl_2 \rightarrow 2NaCl$$

Key term

Chemical reaction A change where a new substance is made and mass is conserved.

9 Write a word equation for the reaction of carbon with oxygen to produce carbon dioxide.
10 Write a word equation for the reaction of sodium with another element to produce sodium oxide.
11 Write a balanced symbol equation for the reaction of potassium with chlorine.
12 Write a balanced symbol equation for the reaction of lithium with a halogen to make lithium bromide.

Evidence for the composition of Earth and its atmosphere

The Earth formed about 4.5 billion years ago and is not the same now as then. Scientists have created a hypothesis that uses models and present day evidence to explain how the Earth has changed. The evidence includes:

▶ Seismic activity from explosions and earthquakes – by studying how seismic waves travel through the Earth and analysing the materials in rocks, scientists deduced the layered structure of the Earth.
▶ Fossil records and rock strata (layers) – these suggest that some continents were connected in the past.
▶ Coastlines of continents – these fit like jigsaw pieces suggesting that Africa and America, for example, were once the same land mass.
▶ Space exploration – evidence from space probes suggests how the early Earth's atmosphere might be like the atmosphere of Venus or Mars today.
▶ Ice cores – samples from glaciers and permafrost in the Arctic contain trapped bubbles of ancient atmosphere that can be analysed; the deeper the ice, the older the air samples.
▶ Modelling – scientists use computer models to show how the Earth's evolution may have occurred.

Key term

Tectonic plates Seven or eight very large slabs of rock that make up the Earth's crust and float on the mantle.

Structure of the Earth

The Earth has a structure made of three layers, each with different properties (Figure 8.4):

▶ The core – the middle of the Earth is mainly iron with some nickel at about 5500 °C. The solid inner core is surrounded by an outer layer of molten metal.
▶ The mantle – consists of semi-molten rock called magma. When this erupts onto the Earth's surface, we call it lava.
▶ The crust – the solid part of the Earth that we live on and up to 60 km thick. It is not one complete layer, but has sections called tectonic plates.

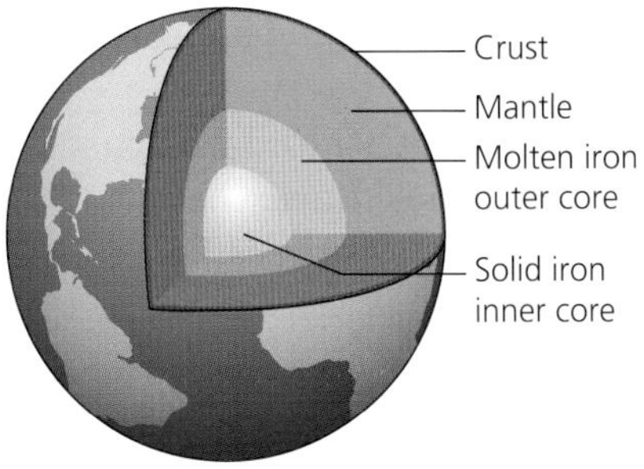

Crust
Mantle
Molten iron outer core
Solid iron inner core

Figure 8.4 The Earth is made up of three main layers

▶ Plate tectonics

The idea that the Earth's landmasses change over time was proposed by Alfred Wegener in the early 1900s in his theory of continental drift. He suggested that all landmasses were initially together in a giant super continent called Pangea that split into two: Gondwanaland and Laurasia. These land masses kept moving, splitting up into the present-day continents.

Wegener also suggested that the plates at the edge of continents moved into each other, causing the surface to buckle and form mountain chains. Before Wegener devised this idea, scientists believed that mountainous 'wrinkles' were formed as the Earth's crust cooled and solidified.

It took nearly half a century for Wegener's hypothesis to be accepted because there was no evidence or theory for how the Earth's plates moved. Wegener only had secondary evidence in the coastlines of continents, and rock strata and fossils. In the 1960s evidence of convection currents in the Earth's mantle was observed and, together with evidence of changes in the Earth's magnetic field, a mechanism for plate movement was established. The theory of continental drift was finally accepted and refined into plate tectonics.

The exact mechanism for the movement of the plates is still not fully understood. Many scientists now believe that slab pull – when older, denser tectonic plates sink into the mantle, dragging along newer, less dense plates   causes the movement of the plates, not the convection currents in the mantle. Scientific theories develop and adapt over time as new evidence is collected.

Plate boundaries

The overall dimensions of the Earth are not changing. So, where new plates are forming, the same volume of old plates must be removed. This happens at plate boundaries. There are three types of plate boundaries (Figure 8.5):

Convergent – the plates move into each other. The less dense plate rises up over the denser plate, which is forced down into the mantle. Explosive volcanoes form along the plate margin (for example, the Andes Mountains in South America). If the plates are the same density, then both plates are pushed up to form fold mountains like the Himalayas.

Divergent – the plates move apart, for example at a mid-oceanic ridge. This creates a gap for the magma to rise up and solidify into new igneous rock.

Conservative – The plates move past each other and no crust is lost or gained. When conservative boundaries move, they create earthquakes but not volcanoes. (The San Andreas Fault in Southern California is an example.)

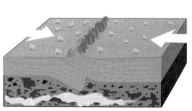

Convergent boundary

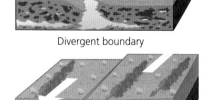

Divergent boundary

Conservative boundary

Figure 8.5 The plates of the Earth are in constant motion and where they meet is called a plate boundary

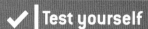

✔ | Test yourself

13 State the main element found in the Earth's core.
14 Which is the thickest layer of the Earth? Look at Figure 8.4.
15 What do scientists think causes the tectonic plates to move?
16 Along which type of plate boundary are mountains produced?

Development of the atmosphere

The Earth's atmosphere changed as the planet cooled and life began. It is difficult to be certain how the atmosphere developed, but scientists have collected a lot of evidence to suggest that the model below is correct. For example, scientists have found that very old igneous rocks contain iron compounds and not iron oxide. This suggests that these rocks could only have been formed when the oxygen levels in the atmosphere were very low.

▶ Initially, there was almost no atmosphere. As the Earth formed, gases like carbon dioxide, water vapour and ammonia were produced from volcanic eruptions. Scientists think that this early atmosphere similar to the atmospheres of Mars and Venus today.
▶ After 0.5 billion years, the Earth cooled. Water vapour in the atmosphere condensed to make seas and oceans, and some carbon dioxide dissolved into the water.
▶ After 2.7 billion years, life began. Cyanobacteria developed and photosynthesis started, releasing oxygen into the atmosphere and reducing the concentration of carbon dioxide. Ammonia reacted with the oxygen, forming nitrogen and water vapour. As complex plant life developed, even more carbon dioxide was removed from the air. Some entered long-term carbon stores, such as sedimentary rocks (chalk and limestone) and fossil fuels.
▶ In the last 200 million years, the composition of dry air has been stable with nitrogen (78%), oxygen (21%), with argon and other noble gases (0.9%). Carbon dioxide, the most abundant gas in the early atmosphere, is now only 0.04%.

Extracting and processing raw materials

The Earth's crust, atmosphere and oceans provide raw materials. We have to extract and process them so that they can be used:

▶ extraction from the environment, for example, mining
▶ physical processing, such as filtering sea water to remove insoluble materials
▶ chemical processing, like the reactions used to remove a metal from its compound.

The main costs of extracting and processing raw materials are people's labour and energy. It is only profitable to extract and purify a material if the selling price is higher than the cost of production. The higher the demand for a material, the greater the selling price and the larger the profit.

Extracting metal resources from the Earth's crust

Rocks are solid mixtures of different substances. Sometimes rocks contain pure metals, such as the gold found in the Clogau Gold Mine in North Wales; sometimes the metals are part of compounds, such as the lead sulfide in the galena ore rocks at the Minera Lead Mines near Wrexham.

Metal **ores** are minerals or rocks containing enough metal compounds for economical extraction and processing. There are two types of mining for removing ores from the Earth:

▶ **Surface mining** – ores are close to the Earth's surface so the topsoil and rock is removed to make a pit. This technique is more common today than it was in the past.
▶ **Subsurface mining** – involves tunnelling under the Earth's surface and extracting the ore without any surface disturbance. A good example of this is the 3500-year-old Great Orme copper mine under the hillside in Llandudno.

Mining impacts the environment as large amounts of waste rock are deposited in huge spoil heaps. Landslides can happen if the spoil is unstable, endangering lives, and chemicals in the waste can contaminate soil and ground water.

Once a metal ore has been mined, the metal compound has to be extracted and chemically transformed to produce the pure metal. Reduction is used to extract metals from their compounds, and the method depends on the reactivity of the metal:

For metals below carbon in the reactivity series, oxygen is removed by reaction with carbon or carbon monoxide. For more reactive metals or to make very pure samples, electrons are provided to the metal ions in the compound by electrolysis. The metal compound must be molten or dissolved in water (aqueous solution) and a lot of costly electricity is needed.

Metals can be recycled (collected, melted down, purified and recast) an infinite number of times. Recycling metals uses much less energy and does not involve further mining, so is better for the environment.

> **Key term**
>
> Reduction Chemical change where oxygen is removed or electrons are gained.

▶ ## Using the blast furnace to extract iron

Monmouthshire and Glamorgan were some of the first ironworks sites of the Industrial Revolution. South Wales still has all the raw materials to make iron:

▶ hematite (iron ore)
▶ coal to make coke to heat the blast furnace
▶ limestone to remove the impurities.

> **Key term**
>
> Blast furnace Tower where iron ore is reduced using carbon. Hot air is blown in.

Limestone undergoes thermal decomposition to make calcium oxide and carbon dioxide. The calcium oxide reacts with silicon dioxide impurities in the iron ore to make calcium silicate, or slag. This is used in road building and to make breeze blocks for houses. The word equations for these reactions are:

Thermal decomposition:

calcium carbonate → calcium oxide + carbon dioxide

$$CaCO_3(s) \rightarrow CaO(s) + CO_2(g)$$

Neutralisation:

calcium oxide + silicon oxide → calcium silicate

$$CaO(s) + SiO_2(s) \rightarrow CaSiO_3(l)$$

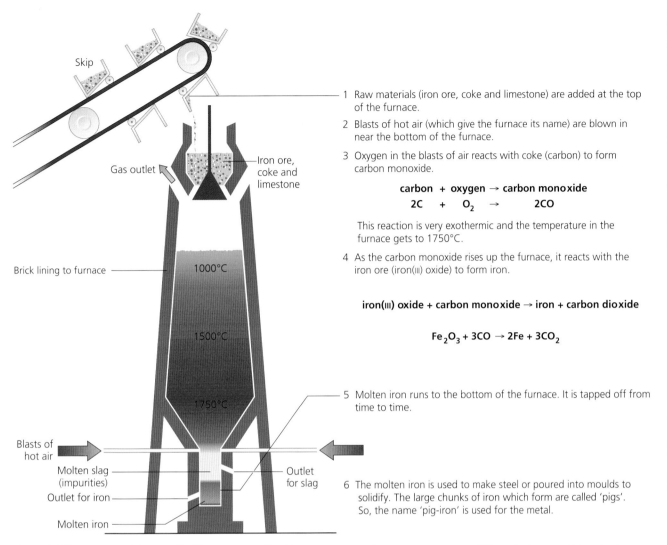

1 Raw materials (iron ore, coke and limestone) are added at the top of the furnace.

2 Blasts of hot air (which give the furnace its name) are blown in near the bottom of the furnace.

3 Oxygen in the blasts of air reacts with coke (carbon) to form carbon monoxide.

carbon + oxygen → carbon monoxide
$$2C + O_2 \rightarrow 2CO$$

This reaction is very exothermic and the temperature in the furnace gets to 1750°C.

4 As the carbon monoxide rises up the furnace, it reacts with the iron ore (iron(III) oxide) to form iron.

iron(III) oxide + carbon monoxide → iron + carbon dioxide

$$Fe_2O_3 + 3CO \rightarrow 2Fe + 3CO_2$$

5 Molten iron runs to the bottom of the furnace. It is tapped off from time to time.

6 The molten iron is used to make steel or poured into moulds to solidify. The large chunks of iron which form are called 'pigs'. So, the name 'pig-iron' is used for the metal.

Figure 8.6 The raw materials are added into the top of the blast furnace and the molten iron is run off from the bottom. Hot air is blown in at the sides

▶ Using electrolysis to extract aluminium

Holyhead in Anglesey was once home to a huge aluminium smelting plant with gigantic electrolytic cells (Figure 8.7).

During the industrial **electrolysis** of aluminium oxide, a mineral called cryolite is added to aluminium oxide to lower the melting point. Once melted, the ions are free to move and the mixture becomes an electrolyte. Large steel vessels are lined with a carbon cathode (negative electrode). Carbon anodes (positive electrodes) are lowered into the molten electrolyte to complete the circuit.

Aluminium ions are attracted to the negative electrodes where they each gain three electrons to be reduced and form neutral aluminium atoms. The molten aluminium metal is run off from the bottom of the electrolytic cell.

The oxide ions are attracted to the anode where they are oxidised by each losing two electrons to become neutral oxygen atoms. These oxygen atoms quickly react with the carbon of the anode to form carbon dioxide gas, so the anodes need to be regularly replaced.

> **Key term**
>
> **Electrolysis** Electricity is used to decompose an ionic substance into simpler substances.

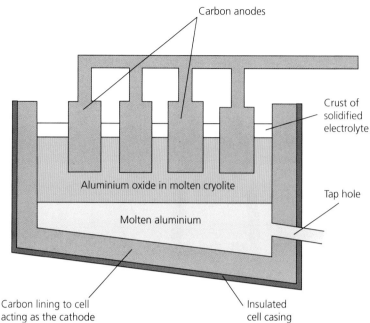

Figure 8.7 Aluminium is extracted from its bauxite ore using electrolysis

Test yourself

20 What is the name of the iron ore used in the blast furnace?
21 What is the main reducing agent in the blast furnace?
22 Give one use of slag.
23 Explain why aluminium cannot be extracted using the blast furnace.
24 Explain how aluminium oxide is reduced in the extraction of aluminium.
25 Explain how aluminium ions are reduced in the extraction of aluminium.

The overall equations for this reaction are:

aluminium oxide → aluminium + oxygen

$$2Al_2O_3 \rightarrow 4Al + 3O_2$$

carbon + oxygen → carbon dioxide

$$C + O_2 \rightarrow CO_2$$

Extracting salt

Sodium chloride, or common salt, can be found as rock salt in the Earth or dissolved in sea water. Rock salt, a mixture of salt and clay, is used directly to grit roads, or can be purified. Salt is an important raw material for industry and is also used in cooking. It can be extracted by:

▶ Deep-shaft mining – a vertical tunnel is made and miners then branch out horizontally to dig out the rock salt deposit.
▶ Solution mining – a shaft is made into the salt deposit and water is pumped in. The soluble compounds, including sodium chloride, dissolve in the water. The solution is pumped to the surface for further processing to obtain dry salt.
▶ Processing sea water – sea water is filtered before evaporating the water to leave white crystals of edible sea salt. Anglesey sea salt is an example of this.

Extracting and using crude oil

Fossil fuels, such as oil, are important for use as chemical feedstocks, as well as fuels. Crude oil is found in pockets under capping rocks. A drill hole can be made into the pocket and oil is pumped to the surface to be processed into lubricants, fuels for vehicles, and plastics.

Crude oil is a complex mixture of hydrocarbons. To be useful, it has to be separated into mixtures of hydrocarbons with similar

Key term

Fossil fuels A finite energy source made from ancient biomass.

boiling points. This happens in oil refineries such as the one at Milford Haven on the Pembrokeshire Coast.

Unfortunately, crude oil is sometimes spilt into the environment when drilling goes wrong, when equipment is being decommissioned or from tankers transporting oil. The oil floats on the surface of water and affects habitats by killing organisms. Cleaning oil spills is very expensive and it can take decades for the habitat to fully recover. Fishing and tourism may be affected.

Fractional distillation

Crude oil is heated to form a vapour and separated in a process called **fractional distillation**. The column is cooler at the top than at the bottom. The hydrocarbons condense at different temperatures depending on their carbon chain lengths and the fractions have specific uses (Figure 8.8).

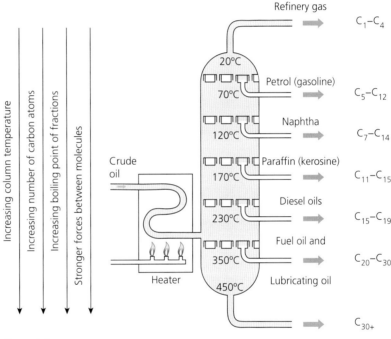

Figure 8.8 Crude oil is separated in a fractionating column

Cracking

Short chain hydrocarbons are in high demand as fuels and those with longer chains are less useful. So the long chain hydrocarbons are **cracked** (broken up) to make more of the shorter (more useful) hydrocarbons. The reactive alkene molecules that are also produced are used to make polymers.

Polymerisation

Polymers are very long molecules made by linking many small repeating units called **monomers**. The alkene molecules that are made from the cracking process have a reactive C=C double bond which allows the molecules to join together. Figure 8.9 shows how monomers join to become a polymer.

The strong bonds that hold polymers together mean they are very durable. Most polymers are not biodegradable and last for hundreds of years, causing pollution and affecting ecosystems. This is a considerable problem in the world's oceans where marine life may die after eating or becoming trapped in plastic pollution.

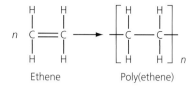

Figure 8.9 Poly(ethene) is a common polymer made from ethene. Ethene is made from the cracking of long chain hydrocarbons from the heavy fractions of crude oil

▶ # Extracting and using shale gas

Shale gas, a fossil fuel, is found in porous rock. There are large deposits of shale gas in Wales. It is extracted by fracking (Figure 8.10). A shaft is drilled into the rock to reach the gas deposit. A high pressure mixture of water, sand and chemicals is pumped into the shaft, pushing the hydrocarbons out. They are collected at the well head for processing.

Shale gas has the same composition as natural gas, so existing equipment and networks are used to process, store and distribute the fuel. Fracking creates jobs and often brings infrastructure improvements, so there are economic benefits.

However, fracking can cause pollution if the waste water is not carefully managed. It can also cause small earthquakes. As shale gas is a finite (non-renewable) resource, fracking is only a medium-term solution to bridge the gap between traditional fossil fuels and renewable, sustainable energy sources.

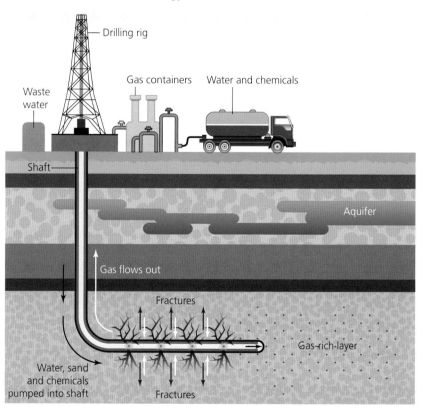

Figure 8.10 Fracking is a new extraction technique which can be used to get shale gas

▶ # Extracting gases from the atmosphere

Air is a mixture of gases with different boiling points. Dry air is compressed and cooled to about −200 °C so that it liquifies; carbon dioxide solidifies and can be removed. The remainder of the liquified air is passed into a fractionating column that is hotter at the bottom and cooler at the top. Nitrogen gas quickly evaporates and flows out of the top of the fractionating column. Liquid oxygen is run off from the bottom of the column.

▶ Extracting biomass

Biofuel is a renewable fuel made from biomass (plants, animal waste, or household waste). For example, ethanol made from the fermentation of crops can be used as an alternative to petrol in cars; in 2021, 10% of petrol in Wales was bioethanol.

Biomass can also be used as a chemical **feedstock** – the manufacture of biodegradable bioplastic uses oil from oil seed rape and corn starch from maize. These crops must be farmed, harvested and processed to yield the feedstocks. Growing biomass crops can push up the price of food and reduce biodiversity, as farmers can choose to grow these over a more diverse variety of food crops.

Unlike traditional plastics, bioplastics are biodegradable – they break down in the environment when microbes use the biomass in the bioplastic as a food source.

> **Key term**
>
> Biomass Dry organic matter made from dead organisms.

> **Key terms**
>
> Bioplastic Biodegradable plastic made from biomass feedstocks.
>
> Biodegradable Capable of being broken down by microorganisms in the environment.

✔ Test yourself

30 Why is shale gas classified as a fossil fuel?

31 What is the raw material used to make carbon dioxide, nitrogen and oxygen?

32 Why are bioplastics better for the environment than traditional polymers?

33 Give one example of a crop that can be used to make bioplastics.

⚙ Specified practical

Preparation of a biopolymer including the effect of a plasticiser

A student prepared two samples of biopolymer from potato starch. She added a plasticiser to one of the samples, to see how it affected the properties of the biopolymer.

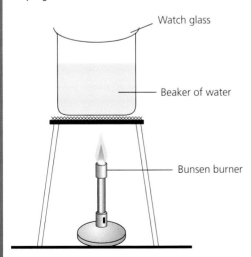

Figure 8.11

Procedure

1 22 cm³ of water, 4 g of potato starch, 3 cm³ of hydrochloric acid and 2 cm³ of propane-1,2,3-triol were poured into a beaker.

2 A watch glass was placed on top of the beaker and the mixture was gently boiled for 15 minutes.

3 The mixture was removed from the heat and sodium hydroxide solution added until the pH was neutral.

4 The mixture was poured into a Petri dish and left to dry producing a biopolymer film.

5 Steps 1–4 were repeated but without the addition of the propane-1,2,3-triol.

6 Compare the two resulting plastics: what do they look like? How easily do they bend?

Questions

1 What safety precautions should the student take?

2 How do the two films compare to each other?

3 Name the plasticiser.

4 Name the polymer.

▶ Product lifespan

Products should be durable, but metals can corrode by oxidation. Corrosion (or rusting) weakens metals and affects their performance. Many metal objects are made from alloys as this reduces their reactivity and rate of corrosion. Metals can also be coated (or painted) to protect them from the environment, stopping or reducing the rate of corrosion.

Electrolysis can be used to electroplate objects with another metal. The item to be electroplated is the negative electrode (cathode) in an aqueous solution of a metal compound. The metal in this solution is deposited on the item in a thin layer. The positive electrode is made from a pure sample of the coating metal. When an electrical current flows, metal atoms from the anode are oxidised and go into solution. The metal ions are attracted to the item at the cathode, where they gain electrons and become neutral metal atoms in the surface coating. Electroplating is commonly used for jewellery and cutlery, as it looks attractive and can protect against wear and corrosion.

Many products are discarded after use, which has an environmental impact. To be more sustainable, we should minimise the materials we use, reuse products if possible, and recycle materials so that we preserve our limited raw materials.

▼ Chapter summary

- The Earth is made of a solid iron core, surrounded by a layer of molten iron, then a layer of liquid rock called magma, and finally a solid crust layer on the outside.
- The crust is made of plates which move. Where they meet, volcanoes and earthquakes can be formed.
- Alfred Wegener first suggested the theory of continental drift.
- The envelope of gas around the Earth is called the atmosphere.
- The atmosphere has developed over time and there are a number of theories to explain how this happened.
- The early atmosphere was likely to be made from volcanic gases and was probably similar to that of Mars or Venus today.
- Oxygen was made when organisms developed and photosynthesised.
- Nitrogen was made when the oxygen reacted with the ammonia in the early atmosphere.
- The carbon dioxide levels reduced due to photosynthesis, dissolving into the newly-formed oceans and the formation of sedimentary rocks and fossil fuels.
- Air can undergo fractional distillation to be separated into carbon dioxide, oxygen and nitrogen.
- Metals are usually found in ores which are mined from the Earth.
- Metal compounds found in ores need to be reduced to extract the metal from its compound.

- Iron is extracted from haematite in the blast furnace by reduction with carbon.
- Aluminium is extracted from bauxite using electrolysis.
- Crude oil is extracted by drilling, then separated using fractional distillation.
- The products of crude oil can undergo cracking to make monomers which can be polymerised to produce polymers.
- Polymers are non-biodegradable, but biopolymers are made from biomass and are more environmentally friendly as they do biodegrade.
- Sodium chloride can be solution mined or deep shaft mined, or extracted from sea water.
- The Periodic Table lists all known elements. The columns are called groups and the rows are called periods.
- Group 1 elements are called the alkali metals and they have increased reactivity as you go down the group.
- Group 7 elements are called the halogens and they have decreased reactivity as you go down the group.
- In a chemical reaction a new substance is made but the total mass remains constant.
- Chemical reactions can be represented in equations where reactants become products.
- Symbol equations must be balanced as no atoms are created or destroyed, they are just rearranged in chemical reactions.

9 Producing useful compounds in the laboratory

Chemicals are important for our everyday lives. We need to find economical and sustainable ways to make and purify substances, with minimum impact on the environment. Chemists develop the best methods to make substances. They work closely with engineers to upscale laboratory batch production into full-scale manufacturing processes.

Acids, bases and alkalis

The Industrial Revolution in Wales started in the lower Swansea Valley due to its plentiful supply of metal ores, coal and water. Sulfuric acid was manufactured for use in the metal extraction industry.

Acids are substances that can dissolve in water and release hydrogen ions in solution (H^+(aq)). Acids react with chemicals called bases. **Bases** include metal compounds such as oxides and metal carbonates.

Some bases, like ammonia, dissolve in water. Soluble bases are called **alkalis** and these release hydroxide ions (OH^-(aq)) into solution.

Acid–base **indicators** can be used to determine whether a liquid is an acid or base. These are substances that have one colour in acid solutions and a different colour in alkali solutions. Table 9.1 shows the colours of common indicators in acids and bases.

Table 9.1 Shows the colours that common indicators go in acid or alkali solution

Indicator	Colour in acid	Colour in alkali
Litmus	Red	Blue
Methyl orange	Red	Yellow
Phenolphthalein	Colourless	Pink

The pH scale

The **pH scale** is a measure of how acidic a substance is by measuring the concentration of hydrogen ions in solution.

Figure 9.1 shows the pH scale:

- pH < 7 are acidic solutions.
- pH = 7 are neutral solutions.
- pH > 7 are alkaline solutions.

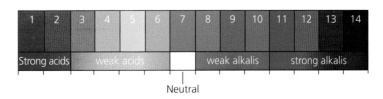

Figure 9.1 The pH scale shows the acidity of solutions

Key terms

Acid A soluble substance that releases H^+(aq) in solution.
Base A substance that reacts with an acid.
Alkali A soluble base that releases OH^-(aq) in solution.

✔ Test yourself

1 What ion does an acid release into solution?
2 How are bases similar to alkalis?
3 How are bases different to alkalis?
4 What colour is litmus when added to hydrochloric acid?
5 What colour is phenolphthalein when added to sodium hydroxide?

Key term

pH scale A measure of acidity of a solution on a logarithmic scale from 0 to 14.

The pH of a solution can be measured with universal indicator. This is a mixture of different dyes which change colour at different pH values. The universal indicator can be a solution which can be added to a substance or on a piece of absorbent paper.

To measure the pH of a solution:

▶ Add universal indicator.
▶ Compare the colour with the colour chart and read off the pH value. Figure 9.1 shows the colours of the Universal Indicator scale.

It can be difficult to interpret the colours on the pH scale. For more accurate results, a pH probe is used (Figure 9.3). It is important to calibrate the pH probe so the results are accurate.

Figure 9.2 A pH probe measures the pH by conductivity and a data logger records the information

Key terms

Neutralisation A chemical reaction between an acid and a base to make a salt.
Salt A neutral ionic compound produced from a neutralisation reaction.

> ✔ **Test yourself**
>
> 6 How would you classify a solution with a pH of 7.00001?
> 7 What does the pH scale measure?
> 8 Which acid–base indicator can be used to determine the pH of a solution?
> 9 Why is using a pH probe more reliable than using an indicator to determine the pH of a solution?

Salts

When an acid reacts with a base, a chemical reaction happens. We know this because new substances are formed. **Neutralisation** is the name of the process when an acid and a base react together to form a **salt**. Salts are made when the positive ion from the base takes the place of the hydrogen ion in the acid. Table 9.2 shows the relationship between the acid and the salt using sodium salts as the example.

Table 9.2 Shows the names of salts produced by different acids

Acid	Name in salt	Example of sodium salt
Hydrochloric acid, HCl	chloride	Sodium chloride, $NaCl$
Nitric acid, HNO_3	nitrate	Sodium nitrate, $NaNO_3$
Sulfuric acid, H_2SO_4	sulfate	Sodium sulfate, Na_2SO_4

Salts can be insoluble or soluble.
Insoluble salts:

▶ do not dissolve in water,
▶ can be extracted from solution using filtering,
▶ include copper carbonate, for example.

Soluble salts:

▶ dissolve in water,
▶ can be extracted from solution using crystallisation,
▶ include copper sulfate, zinc sulfate, potassium nitrate and ammonium nitrate, for example.

Stage 1 THE REACTION
• React the acid with an insoluble substance (e.g. a metal, metal oxide, metal carbonate, metal hydroxide) to produce the desired salt
• Add this substance until it no longer reacts
• This reaction may need to be heated

↓

Stage 2 FILTER OFF THE EXCESS
• Filter off the left over metal/metal oxide/metal carbonate/metal hydroxide

↓

Stage 3 CRYSTALLISE THE SALT
• Heat the solution to evaporate some water until crystals start to form
• Leave the solution to cool down – more crystals will form
• Filter off the crystals of the salt
• Allow the crystals to dry

Figure 9.3 shows an overview of how to make insoluble salts

✔ | Test yourself

10 How can you determine if a salt is soluble?
11 State the name of the salt produced when sulfuric acid reacts with calcium metal.
12 How can soluble salts be extracted from solution?
13 How can insoluble salts be extracted from water?

Metals and acids

Wales has a rich supply of **metals** and metal compounds in its rocks. There is a long history of mining metals that goes back as far as the Bronze Age. Today, there are no working metal ore mines in Wales, but you can still visit them as a tourist and find out about mining in the past.

Metals above hydrogen in the reactivity series react with acids. When a metal reacts with an acid, a metal salt and hydrogen gas is made. The general word equation for this reaction is:

$$\text{metal} + \text{acid} \rightarrow \text{metal salt} + \text{hydrogen}$$

> **Key terms**
>
> **Metal** An element on the left or centre of the Periodic Table or an alloy.
> **Effervescence** Seeing bubbles and/or hearing fizzing.
> **Exothermic** When energy is given from the system to the surroundings.

Figure 9.4 Magnesium fizzing as it reacts with acid giving off hydrogen gas

When solid magnesium metal reacts with a solution of hydrochloric acid we can observe **effervescence**, as shown in Figure 9.4. The bubbles indicate that a gas has been made. We need to test the gas to identify it. If hydrogen is present in a test tube, a lighted splint held near its mouth ignites with a squeaky pop.

As we know that a gas has been made and this is a new substance, we can conclude that a chemical change has taken place.

This reaction also produces a temperature rise, which we can observe by monitoring the reaction with a thermometer. As the temperature increases, we can conclude that this is an **exothermic** reaction.

Magnesium reacts with hydrochloric acid to form magnesium chloride and hydrogen. The equations for this reaction are:

$$\text{magnesium} + \text{hydrochloric} \rightarrow \text{magnesium} + \text{hydrogen}$$
$$\text{acid} \qquad \text{chloride}$$

$$Mg(s) + 2HCl(aq) \rightarrow MgCl_2(aq) + H_2(g)$$

Magnesium chloride is very soluble and forms a colourless solution, so we cannot easily see this product. But if we were to evaporate off the water, as shown in Figure 9.5, the crystals of magnesium chloride would be left behind.

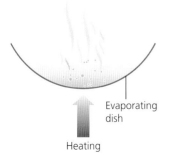

Evaporating dish

Heating

Figure 9.5 The evaporation procedure to extract dissolved salts from a solution

Metal oxides and acids

Metal oxides are bases and so they react with acids to form a metal salt and water. The general word equation for this reaction is:

$$\text{metal oxide} + \text{acid} \rightarrow \text{metal salt} + \text{water}$$

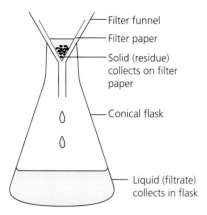

Filter funnel
Filter paper
Solid (residue) collects on filter paper
Conical flask
Liquid (filtrate) collects in flask

Figure 9.6 shows the filtration procedure which can extract insoluble solids from a liquid

When black solid copper(II) oxide reacts with a solution of sulfuric acid, we can observe the solution change colour from colourless to blue. This colour change indicates that a new substance has been made and a chemical reaction has taken place.

The copper salt, copper sulfate, is very soluble and forms the blue solution. To obtain a pure dry sample of this salt, we must first make sure that all the acid has been neutralised. This can be done by adding excess (too much) copper(II) oxide until some of the black solid remains unreacted. The solution is then filtered, as shown in Figure 9.6, to remove the unreacted copper(II) oxide. The copper(II) sulfate solution filtrate is collected. The water is evaporated to leave the solid salt crystals. This is called crystallisation.

Once the blue copper(II) sulfate crystals are collected, they are patted dry with absorbent paper or put into a drying oven to remove any remaining water.

The equations for this reaction are:

copper(II) oxide + sulfuric → copper(II) sulfate + water
acid

$$CuO(s) + 2H_2SO_4(aq) \rightarrow CuSO_4(aq) + H_2O(g)$$

Metal hydroxides and acids

Metal hydroxides are bases and many of them are soluble in water, forming alkaline solutions (alkalis). The general word equation for the reaction of metal hydroxides with acids is:

metal hydroxide + acid → metal salt + water

When a solution of nitric acid is added to a solution of potassium hydroxide, a chemical reaction occurs. It is difficult to observe this reaction because water and a soluble metal salt are formed, which are both colourless. A slight temperature increase indicates that a reaction has taken place. The reaction can be monitored using a pH probe to measure how the pH changes as one solution is added to the other.

Crystallisation can be used to obtain a pure dry sample of the soluble salt. The salt solution is put into an evaporating dish and heated until a hot saturated solution is made, with crystals forming at the edge. The evaporating dish is then removed from the heat and the solution is allowed to cool. The solubility of the dissolved salt reduces as the temperature drops, and more crystals form. Using a spatula, the crystals can be removed and patted dry on absorbent paper or put in a drying oven.

The equations for this reaction are:

potassium hydroxide + nitric acid → potassium + water
nitrate

$$KOH(aq) + HNO_3(aq) \rightarrow KNO_3(aq) + H_2O(g)$$

▶ Metal carbonate and acids

Limestone is a sedimentary rock that contains a lot of calcium carbonate. It is still quarried in Wales today and was processed into other products in locations such as Wrexham's Minera Limeworks. Calcium carbonate is a base. It is used in road building and is added to farmers' fields to increase the pH of the soil and to help improve the yield of crops.

Metal carbonates are bases and react with acids, forming metal salts, water and carbon dioxide gas. The general word equation for this reaction is:

metal carbonate + acid → metal + water + carbon dioxide
salt

Sodium carbonate is an alkali and so it can react with acids, either as a solid or in solution. When sodium carbonate is reacted with hydrochloric acid, there is effervescence and an increase in temperature. The gas produced can be collected and tested by passing it through limewater. Carbon dioxide gas causes the limewater to go cloudy because an insoluble white precipitate forms.

The equations for this reaction are:

sodium carbonate + hydrochloric acid →
sodium chloride + water + carbon dioxide

$Na_2CO_3(s) + 2HCl(aq) \rightarrow 2NaCl(aq) + H_2O(l) + CO_2(g)$

▶ Ammonia and acids

At room temperature, ammonia is a very soluble colourless but smelly gas. It readily reacts with acids and can dissolve in water to form an alkaline solution of ammonium hydroxide. The equations for the reaction of ammonia with water are:

ammonia + water → ammonium hydroxide

$NH_3(g) + H_2O(l) \rightarrow NH_4OH(aq)$

Ammonia gas can react directly with acidic gases such as hydrogen chloride. The equations for this reaction are:

ammonia + hydrogen chloride → ammonium chloride

$NH_3(g) + HCl(g) \rightarrow NH_4Cl(s)$

✔ Test yourself

14 Write word and balanced symbol equations for:
  a) Reaction of copper carbonate and sulfuric acid
  b) Reaction of zinc with sulfuric acid
  c) Reaction of calcium with nitric acid
  d) Ammonium hydroxide with hydrochloric acid
15 Explain why you would never make potassium nitrate from the reaction of potassium with nitric acid.

Ammonia solution or ammonium hydroxide reacts readily with an acid to make a salt and water. For example, ammonia solution can react with nitric acid to make ammonium nitrate and water. This reaction has no visible change but is exothermic, and so it can be monitored using a thermometer. Since it is a neutralisation reaction, it can also be monitored using an indicator. The equations for this reaction are:

$$ammonia\ solution + nitric\ acid \rightarrow ammonium\ nitrate + water$$

$$NH_4OH(aq) + HNO_3(aq) \rightarrow NH_4NO_3(aq) + H_2O(l)$$

⚙ Specified practical

Preparation of useful salts

A student prepared a pure dry sample of zinc sulfate from zinc oxide powder and dilute sulfuric acid. The apparatus that she used is shown in Figure 9.7.

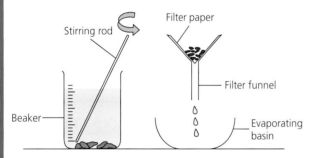

Figure 9.7

Procedure

1. The apparatus was set up as shown in the diagram.
2. 50 cm³ of dilute sulfuric acid was gently heated in a 250 cm³ beaker to 50 °C.
3. 5 g zinc oxide powder was added and stirred. This was allowed to cool.
4. The cooled mixture was filtered.
5. The filtrate was placed into an evaporating basin and heated until the volume of solution was reduced by half.
6. The evaporating basin was removed and left in a warm place until all the liquid had evaporated.
7. A spatula was used to remove the crystals onto absorbent paper and they were patted dry.

Questions

1. State the purpose of the safety precautions that the student must take.
2. State the purpose of step 4.
3. State the purpose of step 5.
4. Write word and balanced symbol equations for this reaction.

▶ Making copper carbonate

Copper(II) carbonate is an insoluble metal salt. It can be prepared using a precipitation reaction:

▶ Add white calcium carbonate powder to a solution of blue copper(II) chloride.
▶ Solid blue copper(II) carbonate is formed in a colourless solution of calcium chloride.
▶ Copper(II) carbonate can be separated by filtering and collected as the residue.

The equations for this reaction are:

$$calcium(II)\ carbonate + copper(II)\ chloride \rightarrow copper(II)\ carbonate + sodium\ chloride$$

$$CaCO_3(aq) + CuCl_2(aq) \rightarrow CuCO_3(s) + CaCl_2(aq)$$

▶ Choosing a reaction pathway

When making a salt it is important to evaluate all the methods that could be used. Think about:

- ▶ **Risk** – You should use the least hazardous chemicals (Figure 9.8 gives the meanings of hazard symbols). Don't use heat sources unless essential.
- ▶ **Skills** – Do you have the skills to carry out the procedure safely? If not, what training do you need?
- ▶ **Purity** – If you want a very pure salt, use a purification step like filtering. You could test the purity using chromatography or by measuring the melting point.
- ▶ **Yield** – Try to reduce the number of times you transfer substances from one reaction vessel to another in order to minimise loss and maximise yield.
- ▶ **Time** – If you don't have much time, you might need to use a quicker method. You may have to compromise on other aspects such as yield, purity or size of crystals. For example, if you want large crystals, evaporate the solvent slowly, but if time is short, use heat to obtain smaller crystals quickly.

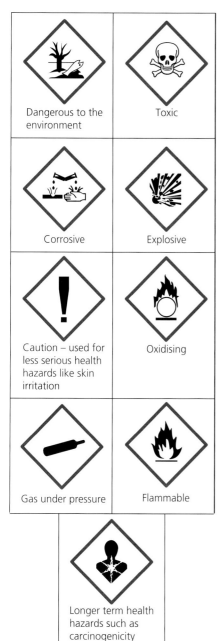

Figure 9.8 The common hazard symbols which can be found on chemicals

⬇ Chapter summary

- Acids have a pH of less than 7 and release $H^+(aq)$ ions into solution.
- Bases react with acids in neutralisation chemical reactions.
- Alkalis are soluble bases with a pH greater than 7 and release $OH^-(aq)$ ions into solution.
- The pH scale is a logarithmic scale to measure the acidity of solutions from 0 to 14.
- Neutral chemicals have a pH of 7.
- Metal oxides, metal hydroxides, metal carbonates and ammonia are bases.
- When a metal reacts with an acid, a salt and hydrogen is produced.
- When a metal oxide or a metal hydroxide reacts with an acid, a salt and water is produced.
- When a metal carbonate reacts with an acid, a salt, water and carbon dioxide is produced.
- When ammonia reacts with an acid, a salt is produced.
- Ammonia dissolves in water to make ammonium hydroxide which is an alkali.
- Salts can be soluble or insoluble.
- Insoluble salts are made in precipitation reactions.
- There are many methods to make soluble salts and chemists must evaluate the methods and choose the one that has the least risk and makes the best yield in the time available.

► Practice exam questions

1 The public water supply used in the home undergoes a number of treatment steps before it is safe to drink.

   a) Name one substance that can be used to disinfect the water. [1]

   b) Explain why distillation is rarely used for the production of drinking water. [2]

   c) Some water is described as hard water. Give the symbols for the two ions that cause water to be hard. [2]

   d) Describe a method that can be used to determine the hardness of water. [6]

2 A safe sustainable water supply is essential for life. Pure water contains only water molecules and has the chemical formula H_2O.

   a) Determine the number of atoms in a water molecule. [1]

   b) Name the elements that make up water. [1]

   c) Explain why water is described as a compound. [2]

   d) Explain why tap water is not pure water. [1]

3 Anglesey Aluminium produced aluminium from 1971 and made 145,000 tonnes of aluminium a year. The raw material was aluminium oxide extracted from bauxite ore.

   a) Write a word equation for the electrolysis of aluminium oxide. [2]

   b) Explain in terms of electrons what happens to the aluminium in the aluminium oxide. [3]

   c) Explain why the carbon anodes need replacing frequently. [2]

4 Raw materials are extracted in different ways.

   a) Identify the correct extraction method for each raw material by drawing a line. [4]

Raw material	Extraction method
Nitrogen	Solution mining
Shale gas	Fractional distillation
Salt	Surface mining
Haematite	Fracking

   b) Crude oil is a complex mixture of hydrocarbons. Name the process used to separate crude oil. [1]

   c) Name the process used to make the starting material for a polymer. [1]

5 Zinc is an element and an essential mineral in the human diet. Sometimes people do not eat enough zinc compounds in their diet and they can take supplements to prevent illness. Zinc can be made into soluble zinc sulfate which can be used in additives and it is easily absorbed by the body.

   a) On a small scale, zinc sulfate can be made by placing zinc metal into acid. State the observations that you would make. [3]

   b) Name the acid which can be used to make zinc sulfate from zinc. [1]

   c) Describe a method to extract a pure dry sample of the zinc sulfate from the reaction mixture. [3]

6 Pembroke Power Plant uses natural gas to generate 2181 MW of power. To prevent any acidic gases from polluting the atmosphere, they are removed before the waste gases are released into the atmosphere.

   a) State the name of the type of substance that could be used to remove the waste gases. [1]

   b) One waste acidic gas is nitrogen dioxide which can dissolve in water to make an acid. State the name of the acid. [1]

   c) Write a balanced symbol equation for the reaction of calcium oxide (CaO) with sulfuric acid (H_2SO_4). [2]

10 Our place in the Universe

Earth is a tiny object in the vast space of the **Universe**. The Universe contains all space, time, matter and energy. It is so huge and contains so many different objects that it is difficult to visualise. We can use parts of the electromagnetic spectrum to image objects in space to gain a better picture of the scale, structure and behaviour of the Universe.

Observing the Universe

The electromagnetic spectrum

The pictures throughout this chapter were taken with a range of different terrestrial and space-based telescopes and cameras, using different parts of the electromagnetic spectrum.

The Universe is continually bathed in all parts of the spectrum. Very hot, super-massive objects like stars, black holes, neutron stars and galactic centres, produce waves in all different parts of the spectrum. Colder lower-energy objects, like planets and the background space of the Universe, emit lower-energy electromagnetic waves like radio waves, microwaves and infrared. The complete electromagnetic spectrum is shown in Figure 10.1.

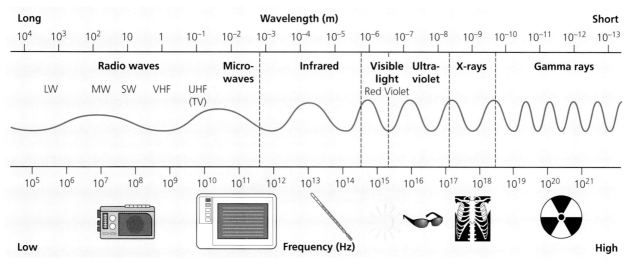

Figure 10.1 The electromagnetic spectrum

Radio waves have the lowest energies, typically around 1×10^{-24} J; visible light has energies around 4×10^{-19} J; and gamma rays have the greatest energies, typically larger than 2×10^{-14} J.

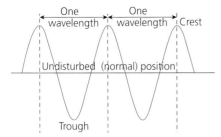

One wavelength · One wavelength · Crest

Undisturbed (normal) position

Trough

Figure 10.2 Wavelength of a transverse wave

Speed, frequency and wavelength

Electromagnetic waves are transverse waves, and the **wavelength**, measured in metres, is the distance a wave takes to repeat itself over one cycle. This is shown in Figure 10.2.

Radio waves have **wavelengths** up to tens of kilometres long, and gamma rays as small as 10^{-13} m (0.000 000 000 000 1 m).

The **frequency** of an electromagnetic wave is the number of waves that pass a point in 1 second. Frequency is measured with a unit called hertz (Hz), where 1 Hz = 1 wave per second. Electromagnetic waves have frequencies varying from 100 000 Hz for radio waves up to 10^{22} Hz (10 000 000 000 000 000 000 000 waves per second) for gamma rays.

The wave speed, frequency and wavelength of electromagnetic spectrum waves are related to each other through the wave equation where:

$$\text{wave speed} = \text{frequency} \times \text{wavelength}$$

★ Worked examples

1 Orange light emitted from the Sun has a wavelength of 6×10^{-7} m and a frequency of 5×10^{14} Hz. Calculate the wave speed of the orange light.
2 The James Webb infrared space telescope's maximum observable wavelength is 28.5×10^{-6} m. If the wave speed of electromagnetic waves is 3×10^{8} m/s, use the equation:

$$\text{frequency} = \frac{\text{wave speed}}{\text{wavelength}}$$

to calculate the frequency of these infrared waves.

Answers

1 $\text{wave speed} = \text{frequency} \times \text{wavelength} = 5 \times 10^{14}\,\text{Hz} \times 6 \times 10^{-7}\,\text{m}$
$= 3 \times 10^{8}\,\text{m/s}$

2 $\text{frequency} = \dfrac{\text{wave speed}}{\text{wavelength}} = \dfrac{3 \times 10^{8}\,\text{m/s}}{28.5 \times 10^{-6}\,\text{m}} = 1.05 \times 10^{13}\,\text{Hz}$

✔ Test yourself

1 State the two missing parts of the electromagnetic spectrum, X and Y, from the table below:

radio waves	X	infrared	visible light	Y	X-rays	gamma rays

2 Which part of the electromagnetic spectrum has:
   a) the longest wavelength
   b) the highest frequency
   c) the lowest energy.
3 Calculate the wave speed of the ultraviolet waves emitted by the Sun with a wavelength of 15×10^{-9} m, and a frequency of 20×10^{15} Hz.

Imaging the Universe

All parts of the electromagnetic spectrum are used to observe the Universe. Table 10.1 summarises the key features.

Table 10.1 Key features of the Universe, observed using different parts of the electromagnetic spectrum

Part of the spectrum	Earth-based or satellite observatory (and example)	Example objects imaged
Radio waves	Earth ATCA radio telescopes in New South Wales	Stars, comets, planets and galaxies. Radio signals are particularly useful for studying relatively low-energy and low-temperature objects, such as the gas clouds produced by exploding supernovae as new stars are forming.
Microwaves	Mostly satellite NASA's Cosmic Background Explorer	The Sun and the Cosmic Microwave Background Radiation (CMBR) – the remnant of the radiation left after The Big Bang.
Infrared	Satellite The James Webb Telescope	Infrared can pass through thick dust clouds in space, so infrared telescopes are particularly good at observing **star-forming regions, planets** and the **centre of galaxies**.
Visible light	Earth or satellite The Hubble Space Telescope	The Sun, planets, moons and galaxies. A selection of images of the planets, taken by the HST is shown in Figure 10.15 below.
Ultraviolet	Satellite The Extreme Ultraviolet Explorer	The Sun, galaxies and planets. Ultraviolet (UV) radiation is produced in large quantities by hot, highly energetic objects.
X-rays/gamma rays	Satellite The Chandra X-ray Observatory	Stars and black holes. X-rays (and gamma rays) are produced as gravity sucks matter into the most extreme objects in the Universe, such as black holes.

Key terms

Galaxy A distant collection of stars in space, orbiting around a common centre of gravity (usually a massive black hole).

Supernova The huge explosion that happens when a massive giant star runs out of nuclear fusion material and implodes (collapses in on itself).

✔ Test yourself

4 Which parts of the electromagnetic spectrum can only be observed from space using orbiting satellite observatories?
5 Which part of the electromagnetic spectrum is used to image the material being sucked into a black hole?
6 Why are infrared waves really useful for imaging hot young stars?

Distances and structures of the Universe

Our Solar System

Relative units – comparing distances and masses in the Solar System

Measuring distances and masses in space is tricky. The numbers are so huge that our common units of distance and mass are far too small. The units used in Table 10.2 are given as relative to the Earth and the Sun, where the mass of the Sun is 1 and the radius of the Earth's orbit is 1.

Table 10.2 Some relative values with their actual values in SI units

Relative unit		Actual value and SI unit
The mass of the Earth, $M_\oplus$ The mean radius of the Earth, $R_\oplus$		$M_\oplus = 6 \times 10^{24}\,\text{kg}$ $R_\oplus = 6\,371\,000\,\text{m} = 6.371 \times 10^6\,\text{m}$
The mean distance from the Earth to the Sun, called 1 astronomical unit (1 AU)		$1\,\text{AU} = 149\,598\,000\,000\,\text{m}$ $(1.5 \times 10^{11}\,\text{m})$
The mass of the Sun, $M_\odot = 1$ solar mass The radius of the Sun, $R_\odot = 1$ solar radius		$M_\odot = 2 \times 10^{30}\,\text{kg} = 333\,333\,M_\oplus$ $R_\odot = 7 \times 10^8\,\text{m} = 0.0046\,\text{AU}$

Table 10.3 shows data on the planets inside the Solar System; the best units to use are relative ones. Distances are usually given in AU and masses in $M_\oplus$.

Table 10.3 The planets in the Solar System

Planet	Symbol	Mean orbit radius (in AU)	Orbital period (in Earth years)	Mean radius (in $R_\oplus$)	Mass (in $M_\oplus$)	Mean surface temperature (in °C)	Day length (in Earth days)
Mercury	☿	0.39	0.24	0.38	0.06	165	59
Venus	♀	0.72	0.62	0.95	0.82	465	243
Earth	⊕	1.0	1.0	1.0	1.0	15	1.0
Mars	♂	1.5	1.9	0.53	0.11	−65	1.1
Jupiter	♃	5.2	12	11	320	−110	0.4
Saturn	♄	9.6	29	9.5	95	−140	0.5
Uranus	♅	19	84	4.0	15	−195	0.7
Neptune	♆	30	170	3.9	17	−200	0.6

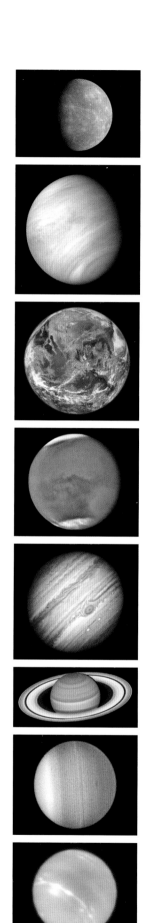

Figure 10.3 The planets of our Solar System (not to scale).

Test yourself

7 What are the actual values (in standard units, metres and kilograms) of the following?
   a) the mean orbit radius of Mercury
   b) the mean radius of Jupiter
   c) the mass of Neptune

8 a) Draw a graph of orbital period (*y*-axis) against average orbit radius (*x*-axis). Make sure you put a title on your graph and label each axis. Draw a best-fit line through your data points.
   b) The best-fit line is not straight. What is the pattern in your data?

▶ How big is our Solar System?

The Sun's gravitational field extends a very long way into space, but there comes a point between the Sun and its nearby stars where the Sun's pull of gravity is less than that of the nearby stars. This happens at about 125 000 AU – over 4000 times further away than the planet Neptune. We use this as the 'edge' of the Solar System. By this definition, our **Solar System** then consists of:

▶ one star (the Sun)
▶ eight planets, (in order: the inner rocky planets, (Mercury; Venus; Earth; and Mars); and the outer gas giant planets, (Jupiter; Saturn; Uranus; and Neptune)
▶ five dwarf planets (Pluto; Ceres; Haumea; Makemake; and Eris)
▶ 214 moons; a **moon** is a natural satellite of a planet or a dwarf planet (like our Moon):
  • Mars has two moons: Phobos and Deimos.
  • Jupiter has 79 moons including its four largest moons: Io; Callisto; Europa; and Ganymede (the largest moon in the Solar System).
  • Saturn has the greatest number of moons, 82, including its largest moon, Titan, which has its own rocky surface and a thick atmosphere, mostly nitrogen and methane.
▶ one asteroid belt (between Mars and Jupiter)
▶ many short-period and long-period comets (such as Halley's comet)
▶ an outer, cloudy halo of small icy objects, called the Oort Cloud.

Figure 10.4 shows what our Solar System looks like as seen from outer space.

The next scale up is our closest group of stars – our galaxy, the Milky Way. Figure 10.5 shows a map of the Milky Way as seen from above and Figure 10.6 is a diagram of the Milky Way drawn from the side.

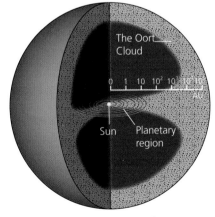

Figure 10.4 The Solar System from outer space

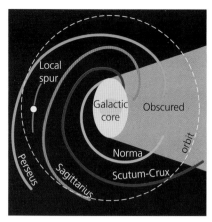

Figure 10.5 The Milky Way from above

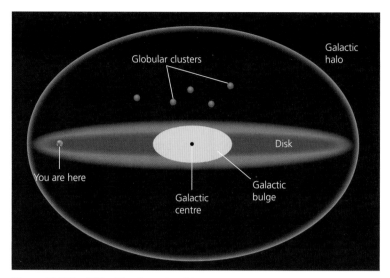

Figure 10.6 The Milky Way from the side

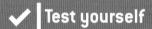

Test yourself

9 What are the main objects in our Solar System?
10 a) What is a 'moon'?
   b) The planets Jupiter and Saturn have the most moons. Why do you think that they have more moons than other planets?
11 a) What is a 'light-year' (l-y)?
   b) How many astronomical units (AU) are there in 1 l-y?

The light-year

The Milky Way is a very big place. The astronomical unit (AU), which we use to compare distances within the Solar System, is too small. The unit we use is the **light-year** (l-y). 1 light-year (1 l-y) is defined as the distance that light travels in 1 year.

Since, the speed of light is 300 000 000 m/s; 1 year contains 365.25 days; each day has 24 hours; each hour has 60 minutes; each minute has 60 seconds:

$$1 \text{ year} = 365.25 \text{ days} \times 24 \text{ h} \times 60 \text{ min} \times 60 \text{ s} = 31\,557\,600 \text{ s}$$

and

$$1 \text{ light-year (l-y)} = 300\,000\,000 \text{ m/s} \times 31\,557\,600 \text{ s}$$
$$= 9\,467\,280\,000\,000\,000 \text{ m}$$

The Solar System is approximately 1.5 l-y across.

The Milky Way galaxy is 100 000 l-y across. Our nearest star, Proxima Centauri, is 4.2 l-y away. The next closest galaxy, Andromeda, is 2.5 million l-y away, and the Universe is about 13.77 billion l-y across.

▶ The Sun

The Sun is a medium sized star and is about 4.5 billion years old. It is the largest object in our Solar System and all the other objects orbit around it. The Sun generates its power output via **nuclear fusion** in its core. Hydrogen nuclei are pushed and fused together at extremely high temperatures (about 15 million °C) and enormous pressures (about 265 billion atm), and some of their mass is converted into radiation energy which is emitted out into space.

As the radiation moves out from the core of the Sun, it travels through its atmosphere. The heat produces convection currents

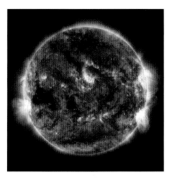

Figure 10.7 Sunspots and solar flares on the surface of the Sun

that cause constant movement of the atmosphere of the Sun. This creates hotter and colder areas on the surface. The colder areas appear as dark 'sunspots' on the surface of the Sun and the hotter areas act as places where solar material is ejected out away from the Sun's surface forming solar flares (Figure 10.7).

Massive solar flares can affect the atmosphere of the Earth, which normally protects us from harmful radiation coming from space. This can cause problems and interference with the Earth's telecommunications systems and can even knock out power grids on Earth, causing widespread disruption.

Analysing the light from other stars and galaxies

Light from stars and galaxies can be separated out into its spectrum of colours (or wavelengths) by a prism. Elements at high temperatures also produce a characteristic spectrum, called an emission spectrum (Figure 10.8).

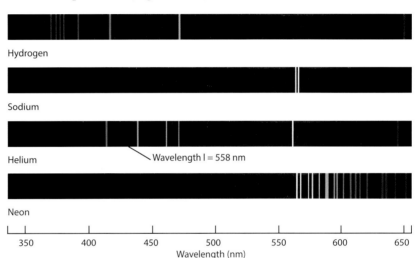

Figure 10.8 Emission spectra of hydrogen, sodium, helium and neon

When light from different elements passes through a gas containing that element (for example, shining hydrogen spectrum light through a mixture of hydrogen and helium gas, like in the atmosphere of a star), the hydrogen gas absorbs the colours of the spectrum, corresponding to hydrogen, forming an **absorption spectrum** (Figure 10.9).

Absorption spectra are a way of identifying different elements on stars far away from Earth and this is called stellar spectroscopy.

> **Key term**
>
> **Absorption spectrum** The pattern of black lines in the spectrum of light from a star that shows the presence of different elements in the atmosphere of the star.

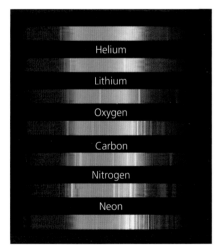

Figure 10.9 Absorption spectra

Red shift

The spectral lines of some stars appear to be 'shifted' towards higher wavelengths, so that they became slightly 'redder'. The patterns stay the same, but each of the different spectral lines moves by the same amount towards the red end of the visible spectrum. This effect is known as red shift (Figure 10.10).

Red shift is caused by the expansion of space and is known as cosmological red shift, modelled in Figure 10.11.

The Universe is represented by the surface of a balloon and a wave of light is drawn on the surface of the balloon before it is inflated. As the balloon is inflated, the surface of the balloon (the Universe) expands and the wave of light is stretched, increasing the wavelength of the light towards the red end of the visible spectrum.

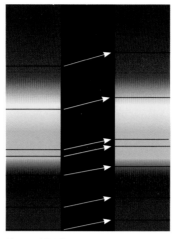

Figure 10.10 Red shift

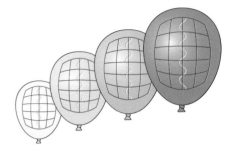

Figure 10.11 Cosmological red shift

✔ **Test yourself**

12 a) What is an 'emission spectrum'?
   b) How is an emission spectrum different from an absorption spectrum?
13 Explain how the emission spectra of elements here on Earth can be used to determine the chemical composition of stars from their spectra.
14 What is 'red shift'?

 # The evolution of the Universe

The Steady-State theory v The Big Bang theory

In 1929, the American astronomer Edwin Hubble published his observations on the expansion of the Universe. His observations, based on 'nearby' galaxies, showed that everyone was moving away from Earth. Modern observations, all show the same pattern – the Universe is expanding. To explain these observations, two theories came into existence in the late 1940s.

The Steady-State theory

The Steady-State theory was proposed by the astronomer Fred Hoyle, and in his theory, although the Universe is expanding, new matter is constantly being created to keep the Universe in a 'steady-state'. In the Steady-State theory, the Universe has no beginning and no end, and always looks the same everywhere.

The Big Bang theory

The Big Bang theory was originally proposed by the physicist Georges Lemaitre in 1931, (although it was Fred Hoyle who first called it 'The Big Bang' theory). Lemaitre proposed that the Universe came into being from a single point (later called the Big Bang), where all the mass, energy, and space of the Universe was created. This point then expanded outwards with time to create the Universe as we see it today.

Evidence for the Big Bang theory from redshift

 In both evolutionary theories, the Universe is expanding. The wavelengths of the light emitted by objects that are moving away from us are increased (or red-shifted) towards the red end of the visible spectrum (as shown in Figure 10.10). The further away the object, the bigger the redshift, so the faster it is moving. Observations show that nearly every object in the visible Universe is red shifting. Hubble used his measurements to conclude that the Universe is expanding. Modern observations confirm his conclusion. The original Hubble dataset is shown in Figure 10.12.

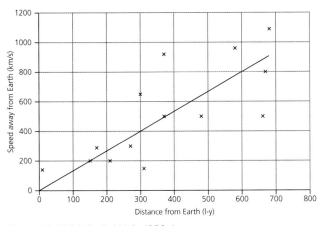

Figure 10.12 Edwin Hubble's 1929 data

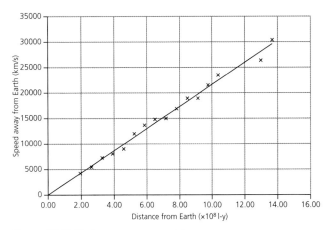

Figure 10.13 Modern supernovae data

The modern data is shown in Figure 10.13. Note that the x-axis scales of the two graphs in Figures 10.12 and 10.13 are very different. Hubble's data would sit in the very bottom left-hand end of the modern supernovae graph!

The first conclusion drawn from this data is that the further away from the Earth, the faster objects are moving away from us. Not only is the Universe getting bigger, but also, the rate that it is getting bigger is increasing, i.e. it is accelerating. The second conclusion that can be drawn, is that if the Universe is expanding, and, at some point in the past it was smaller, in fact, if you extrapolate backwards, at some time in the very distant past, the Universe was just a point – the Big Bang. Working backwards, the Universe was created about 13.77 thousand million years ago (Figure 10.14).

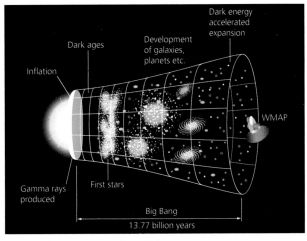

Figure 10.14 Evolution of the Universe

Evidence for the Big Bang theory from Cosmic Microwave Background Radiation

 Whilst the Steady-State theory can also explain the red shift data, the discovery of a microwave background radiation in the Universe could not be explained. Only the Big Bang theory can explain this. The Universe came into being as the result of a huge explosion 13.77 billion years ago, and ever since that time, the Universe has been expanding. At the moment of the Big Bang, huge amounts of energy must have been created in the form of gamma rays. With the expansion of the Universe, the wavelengths of these gamma rays have increased to become microwaves.

The Universe is filled with these microwaves, now called the Cosmic Microwave Background Radiation or CMBR. The WMAP satellite has made a map of the CMBR.

The different colours on the map represent small changes in the intensity of the CMBR. Without these small changes, matter would not have clumped together to form stars and galaxies.

✔ Test yourself

15 State the names of the two competing theories for the evolution of the Universe.

16 According to the Big Bang theory, when was the Universe created?

17 What is the CMBR?

18 What does the WMAP 'map' of the CMBR show us?

⬇ Chapter summary

- The electromagnetic spectrum is a family of waves that have a range of different wavelengths, frequencies and energies, but all travel at the speed of light.
- wave speed $=$ frequency $\times$ wavelength
- there are two theories of the Universe: the Steady-State theory and the Big Bang theory; both have changed over time.
- The Steady-State theory and the Big Bang theory can both explain the red shift expansion of the Universe, but only the Big Bang can explain the Cosmic Microwave Background Radiation (CMBR).
- Images of the Universe can be taken by Earth-based systems and spacecraft, and transmitted to Earth, and then used to study structures in the Universe.
- The light year is the distance light travels in one year.

- Absorption spectra provide us with information about the composition and relative movement of stars and galaxies.
- Our Solar System includes the Sun, planets (rocky planets, gas giants, and dwarf planets), moons, an asteroid belt, comets and the Oort Cloud.
- The radius of the Sun is approximately 200 times bigger than the radius of the Earth and it is about 333 000 times more massive.
- The source of the Sun's energy is nuclear fusion, and there are cooler areas on the surface of the Sun called sunspots; and the Sun emits huge solar flares which can affect telecommunication systems on Earth.
- Data on the planets and other objects in the Solar System can be used to identify patterns and compare objects.

World of life

Ecology is the study of how organisms and their environment interact. Living things are suited to their environment, but some species survive while others do not. Over the course of time, species change, become extinct and new ones appear due to the processes of natural selection and evolution.

To study the ecology of an area we identify species and measure their numbers and distribution.

▶ Biodiversity

Biodiversity is the number and variety of plant and animal species in a particular area. The 'area' can be any size – you could talk about the biodiversity on a seashore, in Wales, or in Europe, for example.

In a thriving ecosystem, many species can survive, so biodiversity is a measure of the health of an ecosystem over time.

▶ Adaptation to the environment

Species become adapted to their environment. Features develop that help the organisms to survive. Two closely related species in different environments will adapt in different ways. Adaptations can be:

▶ Morphological – internal or external structural adaptations of the organism, for example, colour of fur, leg length, petal shape or reduced size of appendix.
▶ Behavioural – the time of day when an animal is active or the type of food it eats (plants have very limited 'behaviours').

The adaptations of a cactus are shown in Figure 11.1. Animals also have adaptations. A common example is the tiger. Tigers are a predator; they have keen eyesight which functions well in the dark when they are most active, as this gives them an advantage over their prey. Their eyes are also at the front of their head to give good depth perception. They have large claws and sharp teeth for catching and killing prey. Their leg muscles are powerful so they can chase prey and their tail aids their balance when they run. Finally, they are striped which breaks up their body outline and means that they are harder to see by their prey.

Organisms that are well adapted to their environment have an advantage when competing for essential resources such as food, and because they are more likely to survive than less well-adapted organisms, they have a greater chance of finding a mate (in the case of animals) and reproducing. Data on the numbers of an organism and how they are distributed in the environment tells us how successful an organism is in an environment. The characteristics of an organism can suggest why they are successful or unsuccessful.

Stem is fleshy and stores water

Leaves are reduced to spines to reduce water loss

Long root system to access water deep underground

Figure 11.1 The cactus has quite extreme adaptations because its desert environment is also extreme

Avoiding adverse environmental conditions

It can be difficult for some animals to survive during severe winters. Low temperatures and shortage of food can mean an increased death rate. Winter occurs regularly every year, so organisms can adapt. Examples of adaptations are:

> **Migration** – the animal population leaves the area and goes to a warmer environment, returning the following spring.
> **Hibernation** – the animals enter a sort of sleep state, in which their **metabolism** is very slow and they do not move (so reducing energy expenditure). They eat a lot before hibernation, so that they have enough food stored in their body to last the winter.

Other examples of adaptations to severe winter conditions are fur growth for insulation, and a fur colour change to white for camouflage in the snow.

Key term

Metabolism All the chemical reactions going on inside a living organism.

→ Activity

Biodiversity in the UK

The graph shows data from the UK Biodiversity Partnership report for 2010.
Over the study period, populations of seabirds increased, water and wetland bird populations were more or less stable, but there was a decline in numbers of woodland and farmland birds.

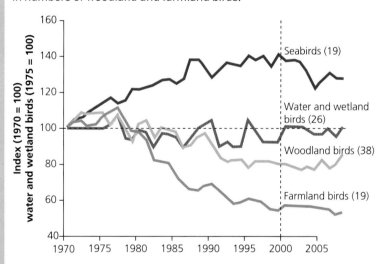

Figure 11.2 Changes in the population of different types of birds in the UK, 1970–2008. The numbers in brackets indicate the number of species monitored

Questions

1. Suggest a possible reason for the decline in woodland birds since 1970.
2. Suggest a possible reason for the decline in farmland birds since 1970.
3. The graph goes up to 2008. If further data had been published for 2010, what do you think they might show? Explain your answer.

Specified practical

Investigation into factors affecting the distribution and abundance of a species

This investigation involves sampling of an area. This has to be done when the area is too big to look at all of it. Whenever a sample is taken:

1 It must be large enough to be representative of the area. In areas that involve different environments (for example, stream, grassland, paths), it is important that each type of area is sampled.
2 It must be taken randomly, to make sure there is no accidental bias.

A group of students estimated the number of daisy plants on a school field. They used quadrats for random sampling, allowing them to estimate the number of daisy plants growing in this habitat.

A simple calculation was used to estimate the total number of daisy species in the entire school field habitat.

Procedure

1 Two 20 m tape measures were laid at right angles along two edges of the survey area.
2 Two 20-sided dice were rolled to decide the coordinates.
3 A 1 m² quadrat was placed where the coordinates met.
4 The number of daisy plants in the quadrat were counted and the results recorded.
5 Steps 2–4 were repeated for at least 25 quadrats.
   The following equation was used to estimate the total number of daisy plants in the field:

Total number of daisy plants in the habitat

$$= \text{total number in samples} \times \frac{\text{total area} \left(\text{m}^2\right)}{\text{total sample area} \left(\text{m}^2\right)}$$

Total area = 400 m² (20 m × 20 m)
Total sample area = number of 1 m² quadrats used

 Worked example

Total number of quadrats = 25

Total sample area = 25 m²

Total number of daisies counted in the sample = 128

Total number of daisy plants in the habitat

$$= 128 \times \frac{400}{25} = 2048$$

Analysing the results

1 Why was rolling dice used, rather than just choosing where to place the quadrats?
2 Why is it important to count at least 25 quadrats?
3 The answer you get from the equation is an estimate. How could you make this estimate more accurate?
4 Why would this method be inappropriate to estimate the number of ladybirds, for example, on the school field?

 Test yourself

1 What is a morphological adaptation?
2 How does hibernation help a species survive in a cold environment?
3 Suggest why, when sampling using quadrats, you would need to take more samples in an area of woodland than on a school playing field.

▶ Classification

To study the huge number of species on Earth, organisms are put into groups. In general terms, plants could be divided into non-flowering and flowering varieties, and animals into vertebrates (with a backbone) and invertebrates (without a backbone). This is not how scientists group living things, however. They have a more complex and detailed system with many more groups, each group having similar and characteristic features. For example, mammals (the group humans belong to) all have hair and feed their young on milk.

Figure 11.3 This is *Oniscus asellus* – also known as a woodlouse, granny grey and a monkey pea, depending on where you come from! Use of these 'local' names could cause confusion, so scientists always use the scientific name

All species have a scientific name and some also have a 'common' name. The scientific name, which always consists of two words, is used by scientists throughout the world. This means that everyone in the scientific community knows which organism is being referred to. Common names vary in different languages (and even in different regions of the same country), so using them could cause confusion. The woodlouse (Figure 11.3), for example, has many different names in different parts of the UK, including monkey pea (Kent), cheeselog (Berkshire), slater (Scotland), granny grey (South Wales) and parson's pig (Isle of Man)! What is more, all of these names are used for all types of woodlouse, but there are 35 different species in the UK.

The system of using two words in a scientific name (the binomial system) was developed by the Swedish botanist, Carl Linnaeus, in the 18th century. The first name is the name of the organism's genus (a group of closely related species) and the second name indicates the species. For example, the big cats all have the same first name, Panthera, but different second names. The lion is *Panthera leo*, the tiger is *Panthera tigris* and the panther is *Panthera pardus*.

The classification of organisms is not just done using external features. Internal anatomy is also important, and genetic sequencing can also be used to group organisms.

Genetic sequencing

Scientists can analyse an organism's DNA and discover which genes are present and their position on the chromosomes. The more similar the genetic sequences are in two organisms, the more closely related they are. This has been used to confirm the classification of some species, and also to show errors and reclassify others. For example, in 2016, giraffes, which were previously thought all to belong to the same species, were reclassified and divided into four different species.

> **✓ Test yourself**
>
> 4 Why do scientists give organisms scientific names?
> 5 Four species of frogs have these scientific names:
>   a) *Rana temporaria*   b) *Pelodytes punctatus*
>   c) *Hyla arboea*   d) *Rana iberica*
>   Which of these species are most closely related? Explain your answer.
> 6 What is gene sequencing used for in classification?

▶ Evolution and natural selection

Variation

Over long periods of time, animal and plant populations change in ways that make them better adapted to their environment. This gradual change is called evolution. If the environment changes significantly, new adaptations may evolve for the new conditions.

Evolution can only happen because populations of living things show variation – the individuals in the population are not identical. The genes of an organism control its characteristics and different sets of genes naturally result in variation; because they are caused by genes, these types of variation can be inherited. These can be regarded as naturally occurring variations.

In humans, the only people with identical genes are identical twins (or other multiple births) because they are formed from the splitting of a single fertilised egg cell. Yet there are variations even between identical twins (Figure 11.4). This is environmental variation and is caused by the influence of the environment – resulting from unplanned life events (such as scars from wounds) and from the individual's personal choices (hair styling, body piercings or tattoos, for example).

Some variations might result from a combination of genetic and environmental factors – height and weight, for example, have genetic components and are affected by diet.

Only naturally occurring variations can result in evolution, as environmental variations cannot be passed on.

Natural selection

The theory of natural selection describes a mechanism by which evolution is thought to occur. It is one of the most famous theories in science and was originated by Charles Darwin.

In the 1830s Darwin went on a five-year scientific voyage of discovery on the ship H.M.S. Beagle. He discovered many new species and noticed different species were variations on a common model, and the variations were linked with the organism's environment or lifestyle.

With another scientist Alfred Russel Wallace, Darwin developed his theory of natural selection to explain the evidence. This was his theory:

▶ Most animals and plants have many more offspring than can possibly survive, therefore the offspring are in a competition for survival. This is the idea of over-production.
▶ The offspring are not all the same; they show variation.
▶ Some varieties are better equipped for survival than others, because they are better suited to the environment. These will be more likely to survive to breed (survival of the fittest).
▶ Those that survive, breed and pass on their heritable characteristics to the next generation (at the time, people did not know about genes).
▶ Over many generations, the best characters become more common and eventually spread to all individuals. The species will have changed (evolved).

Darwin named his theory the theory of natural selection. It has been slightly refined over time but is still accepted as the mechanism for evolution by the majority of scientists.

Figure 11.4 Identical twins have identical genes but there are still some differences between them

Key term

Heritable Capable of being inherited (because it is a result of genes).

7 Which of the following variations in humans is not a 'natural' variation (not caused by the person's genes)?
   a) Eye colour   b) Shape of the ears
   c) Short-sightedness   d) Hair length

8 Why would evolution be impossible if there was no variation in a species?

9 Why are 'beneficial' characteristics more likely to be passed on to the next generation when compared to 'non-beneficial' characteristics?

 Chapter summary

- Biodiversity can be used as a measure of the health of a biological system over time.
- Organisms (plants and animals) are adapted to their environment and this allows them to compete for resources and mates.
- Data (numbers and distribution of organism, characteristics of organism) can be used to investigate the success of an organism in an environment.
- Organisms use strategies such as hibernation and migration to avoid adverse environmental conditions.
- Classification of organisms (plants, animals, microorganisms) is done by grouping organisms that have similar features in a logical way.
- Different groups of organisms can be distinguished according to characteristic features.
- Genetic sequencing is used as a tool to confirm and sometimes reclassify species.
- Classification is not necessarily demonstrated as external features and characteristics.
- There are reasons for the use of scientific names (binomial system developed by Linnaeus) as opposed to 'common' names.
- Natural selection is important as a driving force for evolution.
- Variation occurs naturally due to different genes.
- Individuals with advantageous traits are more likely to be reproductively successful; genes of these individuals are passed on to future generations.

Transfer and recycling of nutrients

Key term

Photosynthesis The process by which plants make glucose using carbon dioxide, water and light energy.

All living things require energy in the form of nutrients in food. Plants make food by **photosynthesis**, and animals eat food. Nutrients are in limited supply and must be constantly recycled.

▶ Food chains and food webs

Energy arrives on our planet in the form of sunlight. This energy passes from organism to organism by means of food chains. Plants change energy in sunlight into stored chemical energy by photosynthesis, so they are the producers and are first in almost all food chains. Plants only manage to capture about 5% of the energy in sunlight.

When plants are eaten by herbivores, some of the energy is passed to the consumers in the food chain. When the herbivore is eaten by a carnivore, the process of energy transfer is repeated. Energy passes from carnivores to scavengers and decomposers, both of which feed on dead organisms. Not all the energy stored by herbivores is available to the carnivores that feed on them; much is used in life processes such as movement, growth, cell repair and reproduction. Some is lost in waste and as heat during respiration.

Consider the food chain through which energy flows when we eat fish such as tuna. Energy from sunlight is first used by plant plankton (microscopic algae). It then passes to animal plankton, then to small fish, then to larger fish, then to tuna and then to us. We are the top carnivores in this food chain.

plant plankton → animal plankton → small fish → large fish → tuna → human

In nature, food chains often interlink, because most organisms eat a lot of different things and are eaten by many different animals. Interlinked food chains are called food webs. Figure 12.1 shows an example, but is an oversimplification of all the feeding relationships that exist in this environment.

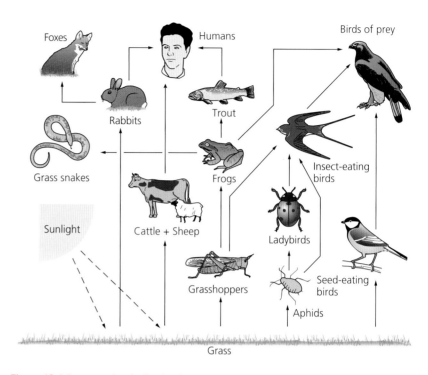

Figure 12.1 An example of a food web

Interdependency of organisms

In any habitat, the living organisms are interdependent (interact with each other in important ways).

Plants may depend on invertebrates and other animals for pollination or dispersal of their seeds. Many plants are insect-pollinated. Animals may disperse seeds attached to their fur or eat fruits containing seeds that later get deposited in the animal's waste, which acts as fertiliser! Animals depend on plants, either directly or indirectly, for food and also for shelter.

Interactions also include those that cause harm, for example, predation, parasitism, disease and competition.

Pyramids of number and biomass

Feeding relationships can be illustrated as pyramids (Figure 12.2 and Figure 12.3). The width of each block in the pyramid indicates the number (or mass) of that type of organism at that feeding level.

These pyramids can tell us more about the energy that is available to organisms living in a measured area or volume. The pyramids can be drawn in different ways:

▶ A **pyramid of numbers** shows the number of organisms per unit area or volume at each feeding level.
▶ A **pyramid of biomass** shows the dry mass of organic material per unit area or volume at each feeding level.

Pyramids of biomass give a more accurate picture than pyramids of numbers. Pyramids of numbers are sometimes not pyramid shaped. Look at Figure 12.4. Here the producers (trees) are much larger than the insects that feed on them. One tree supports thousands of insects, so in a pyramid of numbers the bottom block, representing plants, is narrower than the block for the herbivores. However, a tree weighs a lot more than all the insects feeding on it put together, so a pyramid of biomass will be pyramid shaped.

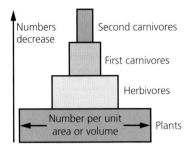

Figure 12.2 A pyramid of number for a grassland food chain

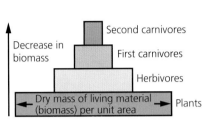

Figure 12.3 A pyramid of biomass for the same grassland food chain shown in Figure 12.2

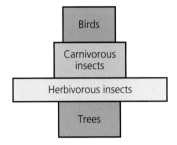

Figure 12.4 Example of a pyramid of numbers that is the 'wrong' shape

✔ Test yourself

1 Do food chains always begin with plants?
2 What is one way that plants may be dependent on animals?
3 What is one way in which animals may be dependent on plants?
4 In what circumstances will a pyramid of numbers not be a pyramid shape?

→ Activity

Calculating the efficiency of energy transfers in a food chain

The amount of energy taken in by organisms at different stages of a food chain was calculated as shown in Table 12.1 below.

Table 12.1 Energy taken in at each stage of a food chain

Stage	Total energy, in kJ
Producers	97 000
First stage consumers	7 000
Second stage consumers	600
Third stage consumers	50

The efficiency of energy transfer at any stage can be calculated as follows:

$$\text{efficiency} = \frac{\text{energy in later stage}}{\text{energy in earlier stage}} \times 100\%$$

Questions

1 Calculate the efficiency of energy transfer for each stage in the food chain.
2 At each stage, the efficiency is quite low. Suggest reasons for this.
3 Using the data, estimate the energy that would be contained in a fourth stage consumer.
4 Suggest why it is unlikely that this food chain could have a fifth stage.

The carbon cycle

Carbon is perhaps the most important element, as all life on our planet is carbon based. There is a fixed supply of carbon on Earth, so it is essential that the carbon on the planet is recycled so the supply is constantly renewed. The way this happens is shown in Figure 12.5.

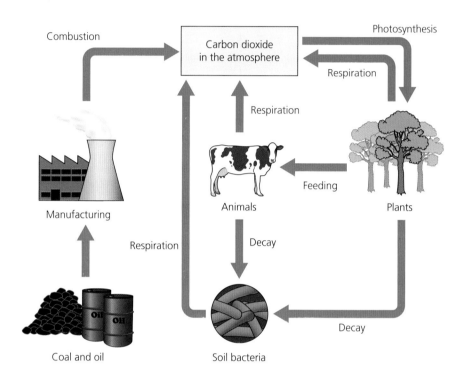

Figure 12.5 The carbon cycle

Carbon dioxide in the air is made into food by green plants in photosynthesis. Animals get their carbon by eating plants (or other animals). The carbon in dead animals and plants is released back into the atmosphere by the process of decay. The bacteria involved release carbon dioxide when they respire. Living animals and plants also respire, and so put carbon dioxide back into the atmosphere.

Fossil fuels were made millions of years ago from the dead bodies of plants and animals. They were not completely decomposed, so the carbon in them was 'locked' into the fossil fuels. When fossil fuels are burnt, this carbon is released as carbon dioxide, adding to the levels in the atmosphere. Humans only started extracting and burning fossil fuels in large quantities in the last 200 years. In addition, humans have cleared huge areas of forest, either for wood supplies or to create new farmland. The trees that were felled previously absorbed a lot of carbon dioxide for photosynthesis. This combination of burning fossil fuels and clearing forests has disrupted the balance of the carbon cycle, leading to an increase in the levels of carbon dioxide in the atmosphere.

The greenhouse effect and global warming

Global warming is one of the major environmental problems facing the world today. It is thought to be caused by the greenhouse effect. The greenhouse effect is summarised in Figure 12.6.

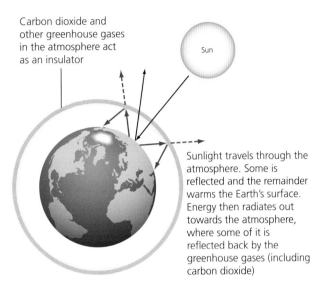

Carbon dioxide and other greenhouse gases in the atmosphere act as an insulator

Sun

Sunlight travels through the atmosphere. Some is reflected and the remainder warms the Earth's surface. Energy then radiates out towards the atmosphere, where some of it is reflected back by the greenhouse gases (including carbon dioxide)

Figure 12.6 The greenhouse effect

The greenhouse effect is necessary to maintain temperatures on Earth that can sustain life and has always existed. However, the 'enhanced' greenhouse effect caused by human activity is having a significant impact on climate and agriculture, and causing polar ice sheets to melt, leading to rising sea levels.

The visible light from the Sun passes through atmosphere. Some is absorbed but most of it reaches the surface of the Earth. A portion of this visible light is reflected into space from the surfaces of oceans, but some is absorbed by the land. The absorbed radiation is then re-emitted back up into the atmosphere, at a much longer wavelength (as infrared radiation). This infrared radiation is absorbed in the atmosphere by a layer of carbon dioxide and other greenhouse gases (mainly methane and water vapour). The increase in greenhouse gases (mostly carbon dioxide from burning fossil fuels) means that more energy is absorbed and less escapes back into space, causing global warming.

> **Key term**
>
> Infrared radiation Electromagnetic radiation that we can feel as heat.

Possible solutions to global warming

Humans can take measures to reduce the problem of increasing carbon dioxide in the atmosphere.

Reducing the use of fossil fuels

Governments, industry, and individuals can:

- use nuclear and renewable power rather than power provided by burning coal, oil and gas.
- recycle or reuse materials as much as possible, so that less fossil fuels are used to make replacements.

- develop and use more fuel-efficient vehicles.
- reduce energy consumption in homes, for example by insulating efficiently, lowering central heating temperatures, using low-energy light bulbs, not leaving appliances on standby, and using and servicing fuel-efficient boilers.
- use means of mass transportation (trains and buses) rather than personal vehicles or share cars where possible.
- improve fuel emissions. It is possible to remove some of the harmful gases produced by the burning of fossil fuels before they enter the atmosphere. At the moment it is only practical to do this on a large scale, such as in power plants.

Carbon capture

Carbon capture can reduce the carbon dioxide emissions from power stations by around 90%. It is a three-step process:

1 Capturing the CO_2 from power plants and other industrial sources.
2 Transporting it, usually via pipelines, to storage points.
3 Storing it safely in geological sites such as depleted oil and gas fields.

▶ Recycling of nutrients

Nutrients on the planet are in limited supply, so it is essential that they are recycled. Organisms can be eaten by others who then obtain nutrients, and when organisms die, the nutrients they contain are released when microorganisms break down their bodies during decay. The nutrients can then be taken up by other organisms such as plants. In this way, the processes which remove materials are balanced by those that return them.

Nitrogen, which is essential for growth because it is used to make proteins, is recycled. The nitrogen cycle is shown in Figure 12.7. Nitrogen in the air is changed by **nitrogen-fixing** bacteria in the soil into nitrates, which plants can absorb and use. Nitrogen-fixing bacteria are also found in the roots of one group of plants, the legumes (peas, beans and clover), in special structures called root nodules. The nitrates absorbed by the plants are passed on to animals that eat the plants, and the nitrogen is eventually returned to the soil in animal urine and faeces, and when dead organisms decay. The nitrogen in wastes and decay is in the form of ammonia, which has been produced by the breakdown of proteins and urea. This cannot be used directly by plants. Bacteria in the soil convert the ammonia into nitrates, which are then absorbed by plants. Nitrogen is returned to the air by denitrifying bacteria.

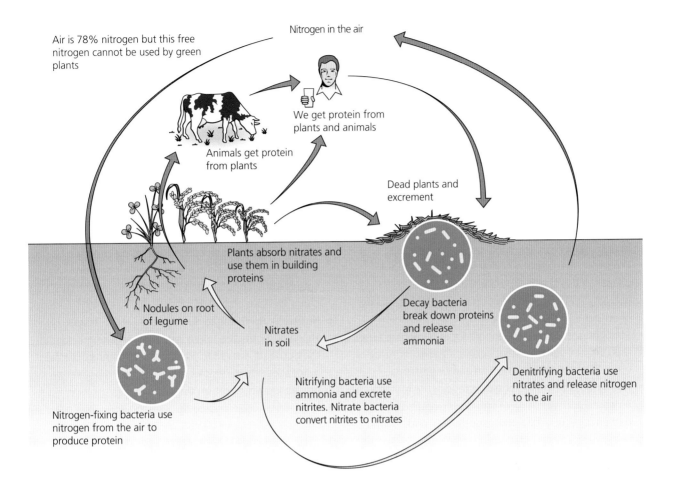

Figure 12.7 The nitrogen cycle

- Food chains and food webs show the transfer of useful energy between organisms.
- Animals can be grouped according to how they feed (e.g. herbivore, carnivore).
- Pyramids of numbers and biomass can be used to indicate the proportions of organisms of different feeding types in an environment.
- Organisms are often interdependent: plants may depend on invertebrates and other animals for pollination, seed dispersal and protection from grazers; animals depend on plants either directly or indirectly for food and shelter.
- Predation, disease and competition are the main causes of death in a population.
- Microorganisms play an important role in the cycling of nutrients.
- Sunlight is the source of energy for most ecosystems.
- Only a small percentage of solar energy is captured by green plants for them to use in photosynthesis.
- There is loss of energy at each stage in the food chain as waste and as heat given off during respiration.
- Carbon is recycled via photosynthesis, food chains, respiration and combustion.
- Bacteria and fungi play a role in the carbon cycle by feeding on waste materials from organisms and dead plants and animals.
- Human activity affects the levels of carbon dioxide in the atmosphere (via burning fossil fuels and clearing forests).

- The greenhouse effect is caused by the Earth absorbing and emitting electromagnetic radiation. Some gases in the atmosphere absorb this radiation, so the planet retains more heat than it otherwise would.
- The greenhouse effect is a natural process needed to maintain life on Earth, but an enhanced greenhouse effect may have significant impact on climate, ice sheets, sea levels and agriculture.
- Proposed solutions to global warming involve reducing human impact on the climate (e.g. cutting dependence on fossil fuels by reducing energy consumption and using alternative non-carbon sources of energy. Carbon capture from industrial chimneys can also help.
- Nutrients are released during decay, e.g. nitrates. These nutrients are taken into other organisms so that the nutrients are recycled; in a stable community the processes which remove materials are balanced by processes which return them.
- Nitrogen is also recycled through the activity of soil bacteria and fungi acting as decomposers. These microorganisms convert proteins and urea into ammonia, which is then converted to nitrates. These are taken up by plant roots and used to make new protein.

H

▶ Practice exam questions

1 The Big Bang theory and Steady State theory are two competing theories of the origin of the Universe. The observations below can be explained by both theories OR only one theory. Copy and complete the table. Tick (✓) the boxes to indicate if the observation can be explained by either, or both theories. [2]

Observation	Explained by The Big Bang theory	Explained by The Steady State theory
The light observed from distant galaxies is red shifted		
Galaxies can be observed in all directions		
The Cosmic Microwave Background Radiation is observed in all directions		
Galaxies that are further away appear to be moving faster than galaxies that are close by.		

2 The diagram below shows the parts of the electromagnetic spectrum.

			Arrow X →			
radio waves	micro-waves	infrared	visible light	ultraviolet	x-rays	gamma rays
			← Arrow Y			

State which arrow represents:

a) Increasing wavelength

b) Increasing frequency

c) Decreasing energy [3]

H 3 The figure plots modern measurements of the recession velocity of galaxies (moving away from the Earth) against their distance away in light years.

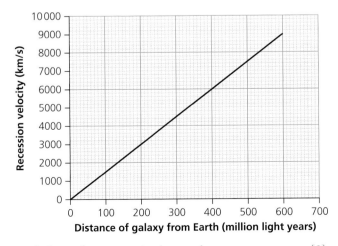

a) State the pattern in the graph. [2]

b) Galaxy UGC 12591 was observed by the Hubble Space Telescope to be 400 million l-y away from Earth. Use the figure to determine the recession velocity of UGC 12591. State the unit with your answer. [2]

c) The Hubble constant, H_0, is defined as:

$$H_0 = \frac{\text{recession velocity}}{\text{distance of galaxy from Earth}}$$

Use this equation, and your answer to b) to determine the recession velocity of the Boötes Galaxy Supercluster 800 million light-years away. [2]

4 Scientists were studying the populations of two British ladybirds – the two-spot ladybird (*Adalia bipunctata*) and the ten-spot ladybird (*Adalia decempunctata*). Despite its name, the two-spot ladybird can have up to 15 spots, either black spots on a red background, or red spots on a black background. *Adalia decempunctata* shows a lot of variation. It can have up to 15 spots and its base colour can be cream, yellow, orange, red, brown, purple or black.

Use the information above to answer the following questions:

a) How do you know that the two-spot ladybird and the ten-spot ladybird are two different species? [1]

b) How can you tell that the two species are closely related? [1]

c) In writing their report, the scientists always used the scientific names of the species. Why is this better than using the common name? [2]

d) Suggest why a description of the colour and number of spots may not be enough to tell which species is which? [2]

e) What causes the variation in colour and number of spots in these species? [1]

5 Malaria is a disease carried by mosquitos. In the middle of the 20th century, the insecticide DDT was used to kill mosquitos in eastern Africa. It was very effective, but a few mosquitos had a natural (genetic) immunity to DDT. The insecticide was used across a wide area, and malaria was almost eradicated. However, by the late 20th century mosquitos and malaria were more common again, and this time DDT was far less effective. Suggest how natural selection can explain these events. [6]

6 Below is an example of a food chain.

grass → mouse → snake → eagle

a) What is the source of energy in all food chains? [1]

b) Suggest why all the energy in the grass is not available to the mouse? [3]

c) Below are some pyramids of numbers. Which one is the most accurate for the example food chain given? [1]

d) Explain why pyramids of biomass are generally more accurate than pyramids of number. [2]

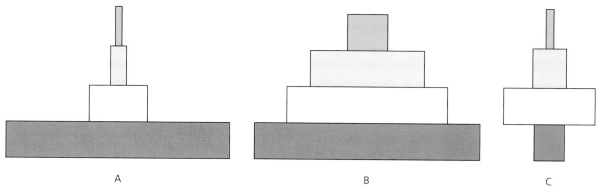

A   B   C

7 The diagram shows the extent of the ice cap at the North Pole in 1979 and 2010.

September 1979

September 2010

a) The effect seen is due to global warming. What is the name of the process that most scientists believe is causing global warming? [1]

b) Why is it important that both diagrams indicate the extent of the ice cap in the same month in both years? [2]

c) Suggest a reason why scientists will continue to measure the ice caps, rather than just relying on data from 1979 and 2010? [2]

d) Why is the shrinking of the ice caps a concern? [2]

e) State one other potential effect of global warming and explain why the stated effect is a problem. [2]

13 Protecting our environment

The environment and its biodiversity are changing due to human impact. Unwanted products affect the world around us. However, we are now making efforts to live more sustainably, treat our waste products more responsibly and maintain biodiversity.

▶ Chemicals and food chains

Living things need small quantities of some metals, but too much is harmful. Some heavy metals, such as lead and mercury, are poisonous even in small quantities. Most heavy metal pollution is caused by industrial processes and mining. Lead pollution from vehicles burning leaded petrol was a problem, but most petrol is now lead-free.

Pesticides are poisonous chemicals that are used to kill agricultural pests, usually by spraying on crops. Some of them take time to break down, so traces of them can be found on fruits and vegetables in shops. Pesticides left in the soil can also be washed into rivers and streams by rain, and pesticide sprays can drift in the air beyond the area being sprayed.

In the UK there are controls on the use of these chemicals; some are banned. The Environment Agency monitors levels of pollutants as there are occasional accidents causing high levels of pollution. Problems arise if these chemicals enter the food chain.

In the 1950s, many people in Minemata, Japan, suddenly started showing the symptoms of mercury poisoning, and 20 died. Mercury was used in a factory on the edge of Minemata bay, but there had been no large spillage. However, the mercury had been absorbed by microscopic plants, which were part of the human food chain (Figure 13.1).

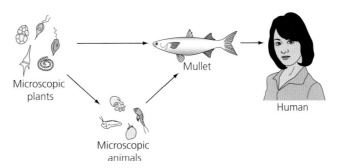

Figure 13.1 Part of the food web around Minemata Bay. The humans ate a variety of other fish and shellfish apart from mullet, but these all ate the microscopic plants and animals

The poisoning happened in this way:

1 Microscopic plants absorbed mercury in the water.
2 Microscopic animals ate large quantities of plants, and mercury built up inside them.
3 Fish ate very large quantities of the microscopic plants and animals, and so mercury built up to even higher levels in them.

4 Fish became poisonous because of the levels of mercury they contained. When humans ate a lot of fish, the mercury levels made them very ill or killed them.

This process of increasing levels of poison moving along a food chain is called **bioaccumulation**. Organisms at the top of any food chain, such as humans, accumulate the highest levels of any poison that enters the chain.

▶ Pollution by sewage and fertilisers

Sewage and fertilisers sometimes get into streams and rivers from farmland, washed from the soil by rain. This starts a process of **eutrophication**, which can kill fish and other animals:

1 Sewage or fertiliser cause an increase in the growth of microscopic plants.
2 Plants have short lives, so the number of dead plants in the water goes up.
3 Bacteria rot the bodies of the plants and the population of bacteria goes up sharply.
4 Bacteria use oxygen for respiration, so the oxygen level of the water goes down.
5 Animals, such as fish, die because there is not enough oxygen in the water.

In addition, microscopic plants may grow so much they form a complete blanket over the surface, blocking the light that the plants at the bottom need to survive.

▶ Plastics and their disposal

Plastics have a wide variety of uses and around 422 million tonnes of plastic products are produced per year. Unfortunately, many of these plastics are only used for a short time. The UK alone produced nearly 50 million tonnes of waste plastic in 2020. Plastic does not break down, so plastic waste remains in the environment indefinitely. Much of this waste ends up in the seas, where it breaks into smaller fragments called microplastics.

The environmental issues with plastics are:

▶ Most plastics are non-biodegradable.
▶ Plastics accumulate in landfill sites and because they do not break down, new landfill sites must be found as others fill up.
▶ Many plastics are single use, so a lot of plastic waste is produced.
▶ Large amounts of plastic reach the oceans, where it can kill animals if they ingest microplastics or get tangled in plastic items.
▶ Toxic chemicals from plastic are in food chains.
▶ Plastic cannot easily be incinerated as toxic gases are given off.
▶ Some plastics are difficult and so expensive to recycle.
▶ Different plastics are recycled in different ways, so the plastic types must be separated before recycling.
▶ Plastics are produced from crude oil, so their production produces greenhouse gases and uses limited fuel resources.

Test yourself

1 How do pesticides end up in rivers and streams?
2 When poisonous chemicals enter an environment, why is this a particular problem for organisms at the top of food chains?
3 When eutrophication occurs, which organisms remove oxygen from the water?

Key term

Biodegradable Capable of being broken down by microorganisms in the environment.

Plastic recycling

There are two main approaches to reducing the amount of plastic in the environment: recycling and reuse. Reuse involves using the plastic item while recycling involves converting the plastic into a new item. Although almost every country has increased plastic recycling, it is still not recycled as much as other materials. The recycling of plastic reduces pressure on landfill sites and decreases the need to produce new plastics from crude oil, so conserving limited natural resources. However, recycling some plastics can be expensive and requires industrial processing. Reuse has none of these disadvantages, but it is not always possible.

▶ Habitat destruction

Key term

Habitat The place where an organism lives.

Over time, living organisms evolve to suit their habitat. If that habitat changes, the organisms may be unable to live there. Humans can bring about huge changes in a habitat very quickly. Animals may lose their food supply, shelter or nesting sites, and plants may be actively cleared. Examples of habitat changes brought about by humans include:

▶ Clearing land for housing, industry, commercial areas, and roads.
▶ Destruction of an area by quarrying.
▶ Dumping of waste materials (landfill, illegal dumping, disposal of nuclear wastes).
▶ The use of land for agriculture, with the associated use of pesticides. This is a major cause of deforestation, for example.
▶ Pollution of various types.

Habitat destruction can sometimes threaten endangered species with extinction but always results in species loss and a reduction in biodiversity.

Biodiversity makes an environment much more interesting, but also makes the ecosystem more stable and better able to resist change. For example, if there is a wide variety of species for a predator to feed on, a reduction in one of those species will not threaten its existence.

Key terms

Biodiversity The variety of living organisms in an area.
Ecosystem A community or group of living organisms together with the habitat in which they live, and the interactions of the living and non-living components of the area.

▶ Measures to ensure sustainability

Reduce the use of plastic

We make and use 20 times more plastic than we did 50 years ago. Efforts to reduce the use of plastics have not been very successful. Plastic consumption grows by about 4% each year in Western Europe (although without the effort to reduce plastic use, the increase would have been larger).

Reuse plastic items

Not all plastic items can be reused, but some plastic items are naturally reused (for example, rigid plastic food containers, kettles and pens). Items that could be reused but are often only used once include plastic bags, drink bottles and cups. There have been efforts to increase reuse of these items by:

- promoting reusable drink bottles and cups
- taxing plastic bags – in 2015 the UK introduced a 5p 'carrier bag tax' and in 2021 we now use 83% fewer single use bags compared to 2014.
- totally banning plastic bags in many parts of Asia and Africa.

→ **Activity**

Plastic recycling

Figure 13.2 shows the recycling rate for plastic packaging in the UK in 2008–2020. The UK's target for plastic recycling is 50% by 2025 and 55% by 2030. Recycling collections by local authorities started in 2007 but were not introduced across the whole country at once.

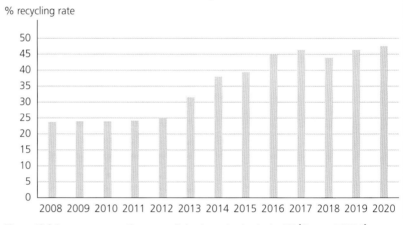

% recycling rate

Figure 13.2 Percentage recycling rates of plastic packaging in the UK (Source: DEFRA)

Questions

1 By how much did the percentage recycling rate rise between 2012 and 2016?
2 Suggest a reason why there was little change in the recycling rate between 2008 and 2011.
3 Suggest a reason why the recycling rate has flattened off between 2017 and 2020.
4 Assuming the current trend continues, do you think the UK will hit the target of 50% recycling of plastics by 2025? Give reasons for your answer.

Biodegradable packaging

What is 'sustainable packaging'? Examples include:

- Replacing single-use plastic bags with paper or fabric ones.
- Using cardboard or recycled paper instead of plastic in pre-packaged items.
- Developing biodegradable 'bio-plastics' from plant materials. Although these plastics are broken down by microorganisms, the process can take many years.
- Selling drinks in cardboard cartons or glass bottles rather than plastic bottles.

✓ **Test yourself**

8 Why is it desirable to maintain biodiversity in an environment?
9 Explain the difference between reuse and recycling.
10 State a problem with recycling plastic.
11 Suggest a reason why the use of cardboard packaging may be better for the environment than biodegradable plastic.

▶ Environmental problems caused by human wastes

Landfill issues

Waste is often disposed of by burial in landfill sites. Most plastics are not biodegradable so landfill sites fill up quickly. Landfill sites cause a variety of environmental problems:

- ▶ Waste materials in landfill sites can generate toxins. These seep through the soil and can enter waterways. This process is made worse by rainfall, which dissolves some toxins and carries them into the soil.
- ▶ Compacting waste can result in lowered oxygen levels. This encourages the growth of anaerobic microorganisms which produce methane, a greenhouse gas.

Household waste

Household waste often contains items which contain environmentally toxic substances:

- ▶ Batteries contain toxic substances, such as sulfuric acid, mercury, nickel, cadmium and lead, as well as lithium, which can cause fires or explosions.
- ▶ Low-energy compact fluorescent light bulbs contain small amounts of highly toxic mercury (LED bulbs are not hazardous).
- ▶ Old mobile phones contain toxins, such as lead, mercury, arsenic, cadmium, chlorine and bromine. If disposed of in landfill, these substances can leak into the groundwater and end up in rivers and streams. Mobile phones also contain lithium batteries.

Sewage

Sewage includes waste from household drains and toilets. The waste water is returned to rivers, but must be treated because it contains potentially dangerous bacteria and can cause pollution (see page 104). In sewage treatment works, solids are allowed to settle and fine waste is broken down into harmless products by bacterial action. The waste is aerated by stirring, because oxygen is needed by bacteria for aerobic respiration. Anaerobic respiration would only give a partial breakdown of the waste, and lack of oxygen can encourage the growth of harmful bacteria. After treatment, the water can be safely discharged into rivers. The solid waste left after treatment can be used as a fertiliser, although there are regulations about how it is used to address possible health concerns.

▶ Biological indicator species

Water pollution can be detected by a fall in the oxygen level or a change in pH, and some pollutants can be measured directly. Scientists can also judge the overall level of pollution using **indicator species**. When some expected species are absent, pollution might be present. Some plants and animals are more tolerant of pollution and these may be observed.

Key term

Indicator species A species with a known tolerance (high or low) to a particular pollutant, which can be used to indicate the level of pollution in an environment.

Lichens are plant-like organisms that grow on rocks, walls and trees. They are used as indicators of air pollution. Figure 13.3 shows lichens that are found in clean air and lichens from areas with different levels of pollution.

Figure 13.3 Lichens from areas with different levels of pollution. The first lichen can only survive in unpolluted air, whereas the final lichen can survive high levels of pollution.

⚙️ | Specified practical

Investigating how indicator species may be used as signs of pollution

Two farms, Mill Farm and Tipton Farm were near the stream, and it was thought that sewage from one or both farms might be getting into the stream. The stream was sampled at five places, labelled A–E on Figure 13.4.

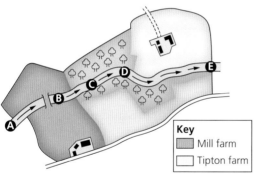

Key
▨ Mill farm
☐ Tipton farm

Figure 13.4 Map of the study area

Procedure

1 Water was collected in a large container.
2 A net was scraped along the floor of the stream and the invertebrates in the net were transferred to the tray.
3 The organisms were identified, and the number of each type of invertebrate was recorded in the table.
4 The contents of the tray were then poured gently back into the stream.
5 Steps 1–4 were repeated at the other locations. The results of the study are shown in the table.

Sample point	Number of each species found per m²							
	Stonefly nymph	Mayfly nymph	Freshwater shrimp	Caddis fly larva	Bloodworm	Water louse	Rat-tailed maggot	Sludge worm
A	11	15	5	12	0	2	0	0
B	0	0	3	4	6	16	12	3
C	0	0	3	8	8	14	2	0
D	0	0	4	10	4	6	0	0
E	0	0	0	4	12	20	2	0

Analysing the results

1 Suggest variables that need to be controlled when taking the samples in different areas.
2 Suggest how these results could have been made more accurate.
3 From the results and information in the table below, what are your conclusions about the levels of pollution in the stream and its likely causes?

Water quality	Species present
Clean water	Stonefly nymph, mayfly nymph
Low level of pollution	Freshwater shrimp, caddis fly larva
Moderate pollution	Water louse, bloodworm
High pollution	Sludge worm, rat-tailed maggot

▶ Sustainable development

Issues of supply and demand

The human population is rising with increasing harmful effects on the environment. More space is needed for housing, roads, industry, mining, agriculture and landfill. Government agencies have an important role in monitoring, protecting and improving the environment. When development is proposed, there is an assessment of environmental impact, including effects on any endangered species. This information helps authorities to decide if the development should be approved, refused, or changed to reduce the effect on biodiversity.

The rising population increases consumption of resources (food, fuels and materials) leading to environmental damage (for example, clearing forests to supply more farmland).

▶ Maintenance of biodiversity

Biodiversity was covered in Chapter 11 World of life. Efforts to maintain biodiversity include the following:

▶ **Sites Of Special Scientific Interest** –areas with legal protection where rare species of plants or animals are found. In Wales, the sites are decided by the Natural Resources Wales organisation.
▶ **captive breeding programmes** – endangered animal species are bred in protected conditions in zoos and wildlife parks, and then released back into the wild.
▶ **national parks** –areas of relatively undeveloped and scenic landscape where building and commercial activities are limited.
▶ **seed banks** – store seeds of endangered plant species so that they can be grown again if they become extinct in the wild. Seeds can germinate into new plants after hundreds of years.

Issues with nature reserves

Local nature reserves are often small and separated. If animals cannot move between patches of suitable habitat, it can make it harder for them to find food or a mate. This reduces the chances of survival of a species. One way of overcoming this problem is to create land corridors connecting reserves together. This allows animals to move between habitats and improves their access to resources. It also helps preserve the genetic diversity of the organisms by reducing inbreeding.

▶ Land reclamation

Industrial areas can fall out of use and landfill sites can become full. Restoring these sites for other uses can be challenging. Buildings may have to be cleared and toxins may be present in the soil. Reclaimed land can be re-planted or may be used for residential or commercial areas (avoiding development of the countryside). In 1961 a reclamation project began in the Lower Swansea Valley, which was a historical centre of metal industry. Industrial ruins and spoil heaps were cleared, trees and metal-tolerant plants were planted, and a new shopping area was built.

Key term

Inbreeding The breeding of individuals which are closely related and so share many similar alleles.

✔ Test yourself

12 Explain how landfill sites might contribute to global warming.
13 What is an indicator species?
14 Explain the difference between a Site of Special Scientific Interest and a national park.
15 Where possible, 'land corridors' are used to connect small nature reserves. Explain the reason for this.

- Heavy metals, from industrial waste and mining, and pesticides can enter the food chain.
- Heavy metals and pesticides can accumulate to toxic levels in animals (bioaccumulation).
- In water polluted by untreated sewage and fertilisers, there is rapid growth of microscopic plants; the death of these results in an increase in the number of the bacteria that break them down, causing a reduction in the dissolved oxygen in the water; animals, including fish, which live in the water may suffocate.
- There are environmental issues relating to the disposal of plastics. They do not biodegrade and so fill up landfill sites. Recycling addresses this and reduces the use of natural resources.
- Habitat destruction results from increased land use for building, mining, landfill and agriculture, causing a loss of species and a reduction in biodiversity.
- Measures to ensure sustainability include reduce, reuse, recycle schemes and the use of biodegradable materials in packaging.
- There are problems associated with unsustainable disposal of waste in landfill sites.
- Environmentally toxic substances are produced by households (sewage, and waste containing toxic substances, such as batteries, low energy light bulbs and old mobile phones); these have an impact on the environment.
- Sewage is treated by the action of microorganisms, so the waste water can enter rivers.
- Environmental monitoring can use living species (lichens to monitor air pollution or invertebrates as water pollution indicators) and non-living indicators (pH and oxygen levels in water).
- There is a need for sustainable development but increased consumption of resources and their continued supply affects the environment.
- Biodiversity can be maintained by captive breeding programmes, seed banks and protected areas.
- Corridors between nature reserves allow movement and prevent isolation between populations of species.
- Reclamation of land previously used for industry and landfill is important for sustainable development.

▶ Practice exam questions

1 Here is a simplified freshwater food chain.

phytoplankton → zooplankton → fish → fish-eating birds

(microscopic plants) → (microscopic animals) → (various types) → (e.g. birds of prey)

DDT, an insecticide, was very widely used in the 1950s and 1960s. Populations of birds of prey declined and they were found to contain high levels of DDT. This did not kill them, but it caused them to produce eggs with very thin shells. These broke, killing the young birds inside.

a) DDT was used on farmland. Suggest how it got into freshwater environments. [1]

b) Suggest why only the fish-eating birds were affected, not the organisms earlier in the food chain. [2]

c) Fertilisers used on farms can also cause damage to freshwater environments due to the process of eutrophication. Explain how this process occurs. [5]

2 Biodegradable municipal waste (BMW) is waste from households which decomposes in landfill to produce methane, a potent greenhouse gas. The graph shows how much BMW was sent to landfill in the countries of the UK between 2010 and 2015.

a) Suggest why the figures for England are much higher than for the other countries. [1]

b) Which country had the biggest reduction in BMW sent to landfill between 2010 and 2015? [1]

c) Suggest why it is a good idea to reduce the BMW sent to landfill. [2]

d) State one way in which other waste, not BMW, can cause environmental damage. [1]

e) Landfill sites are filling up more quickly than in the past, despite the fact that the amount of waste sent to them is decreasing. Suggest a reason for this inconsistency. [3]

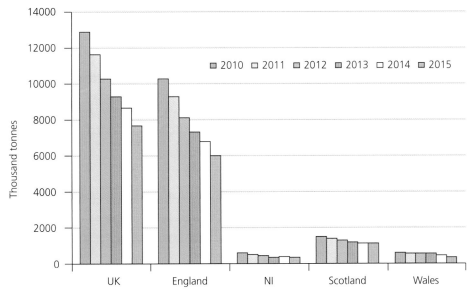

Source: Waste Data Interrogator, Defra statistics

14 Factors affecting human health

Human health is affected by a number of factors including genetics, lifestyle and the environment. We now know about genes and how they affect the way organisms develop. We can also use our knowledge of genes to prevent some diseases. Lifestyle choices have major effects on our health.

▶ The material of inheritance

DNA is the chemical that makes up genes, controlling the structure and function of the body and determining inherited characteristics by controlling the production of proteins.

In the nucleus of a cell, the long DNA molecules are coiled up into structures called **chromosomes**. A gene is a short length of DNA that codes for one protein (Figure 14.1).

DNA is made up of two long chains of alternating sugar and phosphate molecules connected by pairs of bases. This ladder-like structure is twisted to form a 'double helix' (a type of spiral). There are four bases in DNA; adenine (A) bonds weakly to thymine (T), and guanine (G) joins to cytosine (C). The order of these bases along the sugar–phosphate backbone varies in different molecules of DNA. This sequence of bases forms the coded instructions for protein manufacture. It determines the order of the **amino acids** used to make a given protein (Figure 14.2).

Chromosomes in a nucleus match up in pairs containing the same genes, with one member of each pair coming from the father and the other from the mother. Genes have different forms called alleles. For example, the gene for eye colour has blue and brown alleles. Although they contain the same genes, the paired chromosomes may not have the same alleles. The existence of alleles creates variation in a species.

Genetics and genetic terms

You need to know and understand the following specialist terms used in genetics, which will be explained further later in the chapter:

- **Gene** A length of DNA that codes for one protein.
- **Allele** A variety of a gene.
- **Chromosome** A length of DNA that contains many genes, found in the nucleus and visible during cell division.
- **Genotype** The genetic make-up of an individual (for example, BB, Bb, bb).
- **Phenotype** The way the genotype is shown (for example, blue eyes, curly hair, red flowers).
- **Dominant** An allele that shows in the phenotype whenever it is present (shown by a capital letter – for example, **B**).

Gene – a short section of DNA

Chromosome – containing coiled DNA molecules

Cell with nucleus containing chromosomes

Figure 14.1 The structure of a gene in relation to DNA and chromosomes

Key term

Amino acid Chemical group from which proteins are formed.

'Backbone' chains of alternating sugar and phosphate units, twisted into a double helix

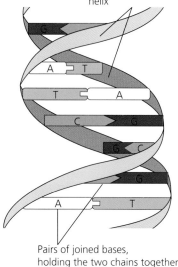

Pairs of joined bases, holding the two chains together

Figure 14.2 The structure of DNA

- **Recessive** An allele that is hidden when a dominant allele is present (shown by a lower case letter – for example, **b**).
- **F1 and F2** Short for first generation (F1) and second generation (F2) in a genetic cross.
- **Homozygous/homozygote** A homozygote contains two identical alleles of the gene – it is homozygous.
- **Heterozygous/heterozygote** A heterozygote contains two different alleles of the gene – it is heterozygous.

▶ Monohybrid inheritance

Monohybrid inheritance is the name given to the inheritance of one gene. There are two copies of each gene in the body, one from each parent. The gametes contain one copy of each gene. The copies are not necessarily the same, as every gene has different versions called alleles. Individuals may have two of the same alleles for a particular gene (homozygous) or two different alleles (heterozygous).

★ | Worked example

In the inheritance of the tall and short characteristics in peas:

- The tall allele (T) is dominant to the short allele (t)
- The dominant allele shows in the phenotype whenever it is present. So, the genotypes TT and Tt both give tall plants.
- The short phenotype is only produced when there are two recessive short alleles (tt).

Show a cross between a homozygous tall plant (TT) and a homozygous short plan (tt), following the inheritance over two generations.

Answer

- Every gamete produced by the tall plant (TT) contains a T allele.
- Every gamete produced by the short plant contains the t allele.
- Whenever pollen from one plant fertilises an egg from the other, the resulting seed will contain one of each allele (Tt). These heterozygous plants are called the F1 generation.
- When an F1 plant produces gametes, half of them will contain the T allele and half the t allele. If F1 pollen fertilises an egg from another F1 plant, there are several possible combinations.
- To work out all these possible combinations, use a Punnett square. Enter the gametes for each individual and then work out the combinations (Figure 14.3).

In the Punnett square, three of the possibilities produce tall plants (TT or Tt) and one produces a short plant (tt). Therefore, tall plants are three times more likely than short plants and, if a lot of offspring are produced, the ratio of tall:short plants would be approximately 3:1. ➡

Key term

Punnett square Table which shows the possible crosses of gametes in a genetic cross.

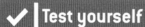

		Male gametes	
		T	t
Female gametes	T	TT	Tt
	t	Tt	tt

Figure 14.3 Punnett square for plant heights

Although you can work out any given cross using a Punnett square, there are two common crosses to remember:

- Aa × Aa gives a 3:1 ratio of dominant:recessive phenotypes.
- Aa × aa gives a 1:1 ratio of dominant:recessive phenotypes.

Test yourself

1 What shape is a DNA molecule?
2 Why are there two copies of every gene in an individual?
3 What is an allele?
4 Explain why an individual may have one allele for a particular characteristic, yet not show that characteristic.

Inherited diseases and chromosome abnormalities

The genes of a species do not remain the same. New alleles and characteristics are constantly appearing. Changes to genes are caused by **mutations**:

- A mutation is a random change in the structure of a gene.
- Mutations are very common. Ionising radiation or certain chemicals can increase the rate of mutation.
- Most mutations cause such small changes that no effect is seen.
- When a change does show, it may be harmful, or occasionally beneficial.
- Mutations in the sex cells (gametes) will be passed on to the next generation. Mutations in body cells will not.

Key term

Gamete Sex cell (egg or sperm in animals, pollen and ovule in plants).

Some mutations can result in an allele that causes a disease. This allele, and therefore the disease, can be inherited. An example is **cystic fibrosis** – the lungs and digestive system of people with this disease become clogged with thick mucus, affecting breathing and digestion, and leading to reduced life expectancy. The cystic fibrosis allele is recessive, so the disease only appears when an individual has the allele for this gene on both chromosomes. Some people have one cystic fibrosis allele and one unaffected allele. People who are heterozygous for the recessive cystic fibrosis trait do not suffer from the disease but can pass it on to their children; they are carriers.

Another example of an inherited disease is Huntington's disease, a condition that affects the brain. It is caused by a dominant allele.

Genetic screening and counselling

It is possible to discover a person's genes by genetic screening, which can reveal whether an adult is a carrier of a genetic disease, or whether a developing foetus has inherited a disease. Screening and counselling help couples who are planning a baby or pregnant women and their partners to make informed decisions about the risk of having an affected child.

→ **Activity**

Genetic counselling and ethical issues

Genetic counselling considers emotive issues and there are no 'right' answers. Discuss these two scenarios, ensuring that you respect the views and concerns of others.

1 Huntington's disease symptoms develop around the age of 30–50. The disease is untreatable and eventually fatal. If a parent has the disease, their child is potentially at risk. Do you think it is a good idea for a 16-year-old with a family history of the disease to have a genetic test?

2 Some people may abort a pregnancy if they know that the foetus has a genetic disorder, even if the child could live with the condition for some time. Discuss the ethical issues around this situation.

Chromosome abnormalities

Some genetic conditions are caused by faults during the formation of gametes that result in an abnormal number of chromosomes. An example is Down's syndrome, which is caused by the presence of three copies of chromosome 21. It happens when a sperm cell or egg cell forms abnormally, so that the gamete has two copies of chromosome 21 rather than the usual one.

▶ Lifestyle and health

Lifestyle and environment can have a major effect on human health. Factors that damage health include:

▶ drugs, including alcohol
▶ obesity
▶ eating disorders
▶ smoking
▶ poor diet
▶ pollutants.

Alcohol

Excessive alcohol consumption can lead to health, behavioural and social problems. Alcohol misuse increases the risk of heart disease, stroke, liver disease, cancers of the liver, bowel, mouth and breast, and pancreatitis, and it can increase risk-taking behaviours. Social problems can include unemployment, divorce, domestic abuse, and homelessness. Alcoholism is an addiction to alcohol.

Obesity

Obesity increases the risk of cardiovascular disease (heart disease and strokes), high blood pressure, some cancers, and type 2 diabetes.

In 2019 61% of the population of Wales was overweight or obese and in the UK, in 2015, the NHS spent an estimated £6.1 billion a year treating obesity and related conditions. If the rate of obesity could be reduced, it would free up funding for the treatment of other conditions.

Obesity is caused by eating too much and taking too little exercise. Poor diet can lead to obesity if it includes too many sugars and fats.

Calculating BMI

Body mass index (BMI) can be used to assess if someone is overweight or obese. BMI is calculated as:

$$\text{BMI} = \frac{\text{mass in kg}}{\left(\text{height in m}\right)^2}$$

A BMI greater than 30 is considered obese. However, BMI can be misleading, particularly for children and athletes. Children and adults have different body composition, so it is best only to compare a child's BMI with others of their age. Athletes build muscle mass so that, according to BMI measurements, many are 'obese'!

★ | **Worked example**

According to the Office for National Statistics, men in the UK have an average mass of 83.6 kg and an average height of 1.75 m. What is the BMI of an average man?

Answer

$$\text{BMI} = \frac{\text{mass in kg}}{\left(\text{height in m}\right)^2} = \frac{83.6}{1.75 \times 1.75} = \frac{83.6}{3.06} = 27.32$$

Anorexia

Anorexia is an eating disorder and mental illness that gives people a distorted body image, thinking they are fat even when they are actually underweight. People with anorexia try to keep their weight down by restricted eating and/or excessive exercising. This effectively leads to starvation. Anorexia can have long-term effects:

Health, fitness and sport

- weak muscles and bones
- problems getting pregnant
- a loss of sex drive
- problems with heart, brain and nervous system
- kidney or bowel problems
- a weak immune system.

► Smoking and health

Smoking is known to damage the lungs. The damaging chemicals in tobacco smoke are:

- **Carcinogens** – chemicals that cause cancer (43 different substances).
- **Tar** – a sticky substance which clogs the bronchioles and alveoli.
- **Nicotine** – an addictive chemical that also directly damages the lungs.
- **Carbon monoxide** – a poisonous gas which makes it more difficult for the red blood cells to carry oxygen.
- Ammonia, formaldehyde, hydrogen cyanide and arsenic, which are present in low amounts.

The effects of smoking on health can be:

- lung cancer
- other cancers (e.g. of the mouth, oesophagus, bladder, kidney and pancreas)
- emphysema (damage to the walls of the alveoli)
- increased risk of heart disease.

Smoking has damaging effects on society, not just the individual smoker. Others can breathe in the smoke and this 'passive smoking' can increase the risk of cancers. In 2019 the cost of smoking-related health issues to the UK Government was approximately £12.6 billion a year, including funding care and treatment, and lost productivity due to worker sickness.

► Diet

There are two important features that determine a healthy or unhealthy diet:

1 The total energy intake.
2 The balance of the different nutrients.

The total recommended energy intake varies for different people, depending on age, weight and height. Data about an individual's lifestyle and medical history is required to make an accurate recommendation.

RI, GDA and RDA

There are daily intake guideline values for nutrients called reference intake (RIs) or guideline daily amounts (GDAs) – the amounts needed to maintain a healthy body.

Another type of guidance is RDA (recommended daily allowance) for vitamins and minerals, which are needed in specific amounts for physical health.

Figure 14.4 Traffic light food label

Food labelling

For the consumer to make dietary decisions, the labelling of food products gives certain information:

- **Food 'traffic lights'**: the Food Standards Agency's system shows green, amber or red coloured labels on the pack to give a quick indication of the balance of nutrients in the food. The label shows whether the product has low, medium or high amounts of fat, saturated fat, sugars and salt.

- **Use by date**: food should not be eaten after its use-by date due to the risk of food poisoning. It is found on high-risk foods, such as fish, meat products, pre-prepared foods, and dairy products. Food that has passed its **best before date** is still safe to eat but may no longer be the best quality. Packaged foods must show either a use by or best before date.

- **List and quantities of ingredients**: all packaged foods which contain more than one ingredient must show a list of ingredients in order of quantity, and the amounts of different nutrient groups, including sugar and salt (usually per 100g).

▶ Salt intake

Salt (sodium chloride) is often added to processed food as a flavouring or preservative. Salt is necessary for correct functioning of our bodies, but too much can have harmful effects.

Too little salt is associated with muscle cramps, dizziness, and electrolyte disturbance. Too much salt in the diet can lead to high blood pressure (leading to an increased risk of heart disease and strokes). The salt requirements of individuals can vary, due to the amount of exercise or physical work they do and how much salt they lose in sweat.

Key terms

Electrolyte A solution containing ions.
Homeostasis The maintenance of a constant internal environment.

> ✔ **Test yourself**
>
> 5 James is a 'carrier' of cystic fibrosis. What does this mean?
> 6 Why are athletes sometimes incorrectly assessed as obese?
> 7 Which substance in tobacco makes it addictive?
> 8 What is a carcinogen?
> 9 Explain the difference between a 'use by' date and a 'best before' date.

▶ Diabetes

Glucose is the body's main source of energy, but it can damage cells if present in high concentrations, so the level is kept within a safe range. This is an example of homeostasis. If the blood glucose level gets too high following a meal, it can be reduced by the hormone insulin which is released by the pancreas into the bloodstream. Insulin converts soluble glucose to an insoluble carbohydrate called glycogen, which is stored in the liver. A second hormone, glucagon, is released from the pancreas if the glucose level gets too low.

In diabetes, the body produces little or no insulin, or fails to respond to any that is produced. If untreated, the blood glucose level becomes dangerously high.

In **type 1 diabetes** (the most common type in young people), the body stops producing insulin. As a result, the blood glucose level goes up and up and the body tries to get rid of the excess glucose in the urine. Doctors diagnose diabetes by the presence of glucose in the urine.

If diabetes is not treated, the blood sugar level becomes so high that the person dies. The condition cannot be cured but it can be managed:

► The person injects insulin (usually before every meal) to replace the natural insulin that is no longer being produced (Figure 14.5).
► The diet must be carefully managed. The patient must eat the right amount of carbohydrate (which is the source of glucose) to match the amount of insulin injected.
► The patient usually tests their blood glucose level several times a day, to make sure the level is not too high or too low.

Type 2 diabetes is more common in older people and happens when the body no longer responds properly to the insulin produced. It is milder and can usually be treated with drugs or dietary control. Type 2 diabetes is associated with obesity and the incidence of type 2 diabetes is increasing dramatically in the UK.

⚙ Specified practical

Investigation of the energy content of food

The energy content of a food is released when it burns. If a burning food is used to heat a known volume of water, the temperature increase can be measured. A simple calculation estimates the amount of energy stored within the food.

Apparatus was set up as shown in Figure 14.5.

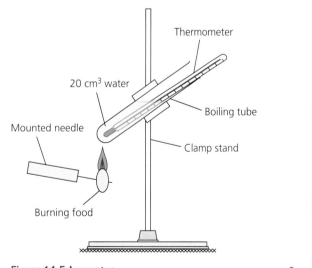

Figure 14.5 Apparatus

1 The temperature of the water was recorded before the start of the experiment.
2 The mass of the food was measured and recorded.
3 The food was set alight and placed in position.
4 The food was held there until it burnt completely.
5 The temperature of the water was recorded again.
6 The experiment was repeated for other foods.

1 Why was the mass of the food measured before it was burnt?
2 When the flame from the food went out, students were asked to try to re-light it. Suggest a reason for this.
3 Although they can be used to compare foods, the actual food energy values are inaccurate. The energy in the food is greatly underestimated in this experiment. Suggest reasons for this.

▶ Epidemiological studies

An epidemiological study looks at the distribution and causes of (or risk factors for) health-related conditions or diseases in a population or area. For example, a study could show possible links between alcohol consumption and liver cancer. Others have shown a connection between diet, obesity and type 2 diabetes. This type of study is necessary because individual cases do not provide valid evidence for or against correlation. For example, the statement 'My grandfather smoked cigarettes for 40 years and never got lung cancer' does not provide evidence that smoking does not increase the risk of lung cancer, it simply establishes that not everyone who smokes automatically gets lung cancer. Most epidemiological studies compare groups of people with different characteristics or lifestyles and investigates if this leads to different outcomes.

These factors increase confidence in the results of epidemiological studies:

▶ The sample sizes must be large enough. Small groups may not be typical of the population.
▶ The profile of samples should match, as far as possible. When comparing two groups, it is easy to ensure an equal proportion of males and females in each group, and a similar age structure. The influence of factors that are not so easily controlled can be overcome using large sample sizes.

It is important to understand that correlation does not mean causation. For instance, between 2000 and 2009, there was a strong correlation in the US between crude oil imports and the consumption of chicken, but clearly eating chicken does not boost oil imports.

> **Key term**
>
> Correlation A connection between two or more things, so that when one of them changes, the other also changes in a predictable way.

▶ Pollutants and health

Heavy metals

We saw that pollution of water by mercury in Japan caused fatalities (page 103). A more widespread problem has been lead pollution.

Lead poisoning, which tends to occur slowly, can be fatal. Lead is found in lead-based paints, which was used in the past. Water pipes were once made of lead so people ingested lead with their drinking water, but these have now been largely replaced. The symptoms of lead poisoning include aggressive behaviour, loss of developmental skills in children, loss of appetite and memory loss, and anaemia (a low number of red blood cells).

Atmospheric pollutants

A variety of air pollutants have known or suspected harmful effects on human health, especially for people who already have lung disease or asthma. These include:

Particles (dust, smoke and smaller particles) from many different sources. If breathed in, they can cause inflammation of the lung tissues.

Acid gases (nitrous oxides, sulfur dioxide) produced mostly by the burning of fossil fuels. These irritate the airways.

Ozone is another irritant gas, which can be formed by the interaction of nitrous oxides and sunlight.

Carbon monoxide gas prevents the uptake of oxygen by the blood and can be fatal if inhaled in large quantities. It can be produced by vehicle emissions, but fatalities often result from poorly maintained domestic boilers if the air supply is restricted.

Test yourself

10 Explain the saying 'Correlation does not equal causation'.
11 What is the difference between type 1 and type 2 diabetes?
12 What effect does insulin have on blood sugar levels?
13 Suggest a reason why lead poisoning is not as common now as it used to be?

Chapter summary

- Chromosomes contain DNA molecules which determine inherited characteristics, and they are found in pairs. Genes are sections of DNA molecules on chromosome pairs that determine inherited characteristics. Different forms of genes called alleles cause variation.
- A DNA molecule consists of two strands coiled to form a double helix, joined by weak bonds between complementary base pairs (A and T; C and G).
- Punnett squares can be used to explain the outcomes of monohybrid crosses.
- The terms genotype, phenotype, recessive, dominant and allele are used in genetics.
- Some alleles can cause inherited diseases (e.g. Huntington's and cystic fibrosis).
- There are implications of genetic screening and subsequent counselling, and ethical problems posed by an individual's prior knowledge of the probability of having a genetic disease.
- New genes result from changes (mutations) in existing genes. Mutations may be harmless, beneficial, or harmful and may be passed on from parents to offspring.
- Chromosome abnormalities in humans can cause genetic conditions, e.g. Down's syndrome.
- There are short-term and long-term impacts of excessive alcohol consumption on the body and society. Addiction can result as a consequence of sustained alcohol consumption.
- Body Mass Index (BMI) can be calculated and used to assess obesity. There are limitations of BMI, particularly for children and athletes.

- Anorexia and obesity have social and economic impacts, and long-term harmful effects. There are adverse health risks associated with obesity, diseases of the cardiovascular system and diabetes.
- Smoking has harmful effects on the body and society.
- Epidemiological studies inform our knowledge of the impact of lifestyle (e.g. smoking, alcohol consumption, diet) on health, but such studies have to follow scientific principles.
- Guideline Daily Amounts (GDA) and Recommended Daily Allowances (RDA) are relevant to a controlled diet.
- Food labelling includes a 'traffic light' system, use by dates, quantities and energy values of nutrients and other components of food, including salt and sugar.
- Insufficient salt intake results in symptoms, such as muscle cramps, dizziness, and electrolyte disturbance. There are risks associated with excessive salt intake (high blood pressure, stroke).
- Diabetes leads to high blood sugar (glucose) level. The presence of glucose in urine is used by doctors to diagnose diabetes.
- Insulin plays a role in the control of blood glucose. Type 1 diabetes is due to insufficient insulin production; in type 2 diabetes body cells do not respond to the insulin that is produced. Type 1 and type 2 diabetes can be controlled.
- Pollutants affect human health, e.g. atmospheric pollutants linked with asthma, heavy metals).

15 Diagnosis and treatment

Being able to image an internal medical problem or injury without surgery leads to much better outcomes for patients. Some medical conditions can be treated using drugs that are tested to ensure that they are safe. Other conditions can be treated using ionising radiation.

▶ Diagnosis

Medical imaging

Medical imaging of structures inside the body uses:

▶ high frequency sound waves, called ultrasound
▶ the highest energy waves of the electromagnetic spectrum (X-rays and gamma rays) (see Figure 10.1)
▶ high intensity magnetic fields, for magnetic resonance imaging (MRI).

X-rays and gamma rays are examples of **ionising radiation**, which can harm or kill living cells. Ultrasound and MRI both use non-ionising radiation.

Ultrasound

The ultrasound waves used in medical imaging have frequencies of between 2 and 20 MHz (2 000 000 Hz to 20 000 000 Hz), giving them a wavelength range of about 0.1 mm to 1 mm inside human tissue. This means that ultrasound is useful for imaging objects within the body that are about this size or bigger. The ultrasound scan of a human baby, inside their mother's womb, is shown in Figure 15.1.

Ultrasound can travel through soft body tissue but is reflected off boundaries between different tissues. The time delay between the ultrasound being sent and received back is measured by the machine. All the time delays are processed by a computer and an image of the body below the ultrasound probe is produced.

Ultrasound is non-ionising to living cells so it is safe for foetal scans. Ultrasound is also used to image joints and to investigate blood flow through the major blood vessels and heart.

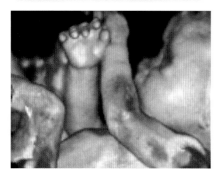

Figure 15.1 A human baby inside their mother's womb

Magnetic Resonance Imaging (MRI)

MRI is another non-ionising form of medical imaging. The patient is placed inside a powerful and varying magnetic field as pulses of radio waves are sent into their body. The radio wave pulses interact with the protons inside the molecules of the body tissue, causing them to emit a different radio pulse back. All the pulses sent back out of the body are combined together and an image of the body is produced (Figure 15.2).

MRI is used on a whole range of body parts in children and adults. It can produce 'sliced' images though the body, so it is useful for imaging soft tissues surrounded by bones, such as the brain and the spinal column.

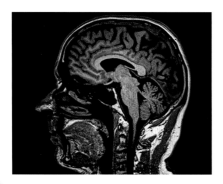

Figure 15.2 An MRI image of the head

X-rays and CAT scans

X-rays are high energy, high frequency, short wavelength electromagnetic waves. X-rays can pass through soft body tissues, but they are absorbed by bones. As a result of this they are ideal for imaging teeth and bone fractures.

In a standard 2-dimensional X-ray, the waves pass through the body and are absorbed at different rates by the different body tissues and bones. The X-rays that pass through are captured by a detector, making an image (Figure 15.3).

The image is a negative – the places where X-rays are absorbed, show up as light areas, whereas places where they pass through are black. Soft tissues, which partially absorb X-rays, show up as grey.

CAT (computerised axial tomography) or CT (computed tomography) scans involve using an X-ray machine that scans around an axis, with the body part to be imaged in the middle.

As the X-ray machine moves around the axis, X-ray images are taken from all angles and the images are combined by a computer to give a very detailed picture (Figure 15.4).

CAT scans are taken in 'slices' that are stacked together to make a 3-dimensional picture. This is very useful for surgeons who need an all-round view of a body part.

CAT scans are useful for measuring and monitoring cancer tumours, imaging damage to bones and internal organs, and checking for strokes.

Gamma cameras

Gamma rays come from the nuclei of radioactive materials. Gamma rays are ionising, so are dangerous to living cells, causing cancers or killing cells. Gamma rays have energies higher than X-rays so they can pass through most materials, including skin, soft tissues and bones. Gamma rays are used to image organs and blood vessels.

A gamma-ray-emitting tracer material is chemically attached to a drug that is absorbed by the part of the body to be imaged. The tracer is injected into the patient's bloodstream which carries it to the target organ. The tracer is absorbed by the organ and emits gamma rays, which are detected by a gamma camera (Figure 15.5). Figure 15.6 shows gamma rays being emitted by two working kidneys.

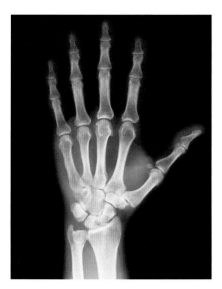

Figure 15.3 An X-ray of the hand

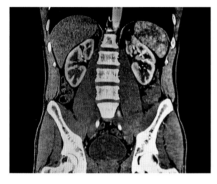

Figure 15.4 A CAT scan of the abdomen

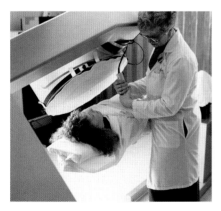

Figure 15.5 A medical gamma camera

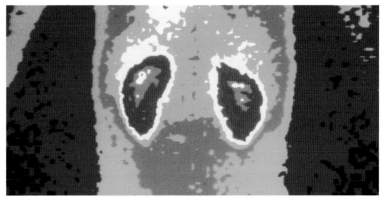

Figure 15.6 Gamma ray image showing two working kidneys

Gamma ray images are useful for monitoring how an organ is working. For example, if part of a kidney is not working, the drug does not attach there and no gamma rays are emitted.

> ✔ **Test yourself**
>
> 1 State the two medical imaging techniques that are non-ionising.
> 2 State the two medical imaging techniques that both use X-rays.
> 3 Which medical imaging technique uses radioactive tracers?
> 4 Which imaging technique would be best to check the growth of a foetus?

▶ Treatment

Drugs

Many common diseases are treated with **drugs**:

- to destroy infectious **pathogens** (antibiotics kill bacteria)
- to treat cancers
- to relieve the **symptoms** of disease, but not kill the pathogen (paracetamol and ibuprofen relieve pain and lower fevers).

As well as having beneficial effects, drugs often have **side effects**. If there are serious side effects, it is important to decide whether the benefit from the drug is worth the side effect.

A good example of a common drug with positive effects and also side effects, is aspirin, a pain relief medicine. Aspirin is also used as a treatment for cardiovascular disease (heart disease), as it reduces the risk of blood clots and heart attacks. A negative side effect is that it can cause bleeding and ulcers in the stomach.

According to some medical studies, in patients who have already had a heart attack or a stroke, daily aspirin prevents approximately 1 in 50 from having further heart problems. However, it causes stomach bleeding in about 1 in 400 patients. So, the benefits of treatment are about eight times higher than the risk of side effect.

Drug testing

When a potential new drug is discovered, it goes through a rigorous and lengthy test procedure before it is released for general use. The development time for a new drug can be as long as 20 years. Drug approval involves several stages:

Preclinical testing

1 A possible drug is identified through research. It is tested using computer models and on human cells grown outside the body in a laboratory. Many drugs fail at this point because they do not work very well or harm cells.
2 The drug is tested on animals, which are monitored for side effects. Animal testing has ethical concerns:
   - For some people, animal testing is acceptable if there are definite human benefits, there is no other way to test and animal suffering is minimised.
   - For other people, it is unacceptable because human benefits cannot always be proven, other testing methods could produce the same outcomes and animals suffer.

> **Key terms**
>
> **Drug** A chemical that alters the way that the body works in a certain way.
> **Pathogen** An organism that causes disease, such as a virus, bacteria, fungi, or parasite.
> **Symptom** A condition of a disease, caused by the disease, e.g., a fever is a symptom of flu.
> **Side effect** The negative effect of a drug on the body.

3 The drug undergoes a **clinical trial**. It is tested on healthy volunteers, initially in very low doses.

4 Further clinical trials are carried out to establish the optimum dose for the drug.

Clinical testing

5 The drug is then trialled on a sample of volunteers who have the disease or condition it is intended to treat, to see if it is more effective than current treatments. The test group receives the drug and the control group receives a **placebo**.

6 If the drug passes all of these tests, it is licensed for general use.

Ionising radiation therapy

Ionising radiation therapy uses alpha (α), beta (β), and gamma (γ) radiation – the products of nuclear decay – as treatments for diseases such as cancers. These are ionising forms of radiation because they can harm or kill living cells.

Radioactivity and nuclear radiation

Radioactivity is a naturally occurring physical phenomenon. The nuclei of some atoms are unstable. To become more stable, they can emit particles or rays of nuclear radiation (Figure 15.7) that take energy away from the atom's nucleus:

▸ alpha (α) particles
▸ beta (β) particles
▸ gamma (γ) rays.

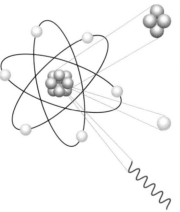

Alpha (α) radiation
These are particles, not rays, and travel at about 10% of the speed of light. An α particle is identical to a helium nucleus, consisting of 2 protons and 2 neutrons joined together.

Beta (β) radiation
These are fast-moving electrons that come from the nucleus. They travel at about 50% of the speed of light.

Gamma (γ) radiation
This is an electromagnetic wave. It travels at the speed of light (3×10^8 m/s). It has very high energy.

Figure 15.7 Alpha, beta and gamma radiation

▸ Alpha particles are helium nuclei. They are the most ionising and the least penetrating type of radiation – they are absorbed by a thin sheet of paper or by the skin.

▸ Beta particles are high-energy (velocity) electrons ejected from the nucleus of a decaying atom. They have medium ionising ability and are absorbed by a few millimetres of aluminium or Perspex.

▸ Gamma rays are high-energy electromagnetic waves. They are the least ionising, but are the most penetrating, able to travel through several centimetres of lead.

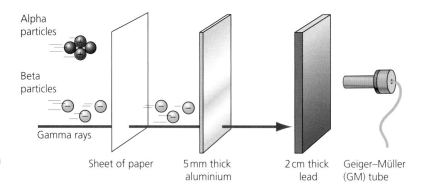

Figure 15.8 The penetration of alpha, beta and gamma radiation

The energy emitted as alpha, beta or gamma radiation moves out and away from the atoms, killing or harming any human tissue in the way. This property can be used to treat cancer cells that are growing out of control, because these cells are more susceptible to damage from ionising radiation and die or reproduce more slowly. A side effect of this therapy is that some healthy cells are also affected by the treatment. As with drug testing, the benefits of ionising radiation therapy must be considered compared to the risk of possible side effects.

Half-life

Radioactive decay is a random event, governed by the laws of probability. With a collection of 120 radioactive atoms, you cannot say which atoms will decay in a given time, in the same way that you cannot say which of 120 dice will throw a three. What you can say is, if you throw 120 dice, the probability is that one in six will show a three, so you expect 20 will show a three.

Any atom of a radioactive isotope has the same chance as any other of decaying. This can be expressed as a **half-life** – the time that it takes for half the atoms in any sample to decay, so the activity is halved. For any one type of atom, the half-life is constant. Radioactive isotopes with very long half-lives remain radioactive for a very long time, whereas isotopes with very short half-lives only remain radioactive for fractions of seconds.

The unit of radioactive activity is the becquerel, Bq, named after Henri Becquerel, who discovered radioactivity in 1896. An activity of 1 Bq is equivalent to 1 radioactive decay per second, which is quite a low value. A 0.5 g iridium-192 wire (a beta emitter used to treat cancerous skin tumours) has a total activity of 160 000 000 000 000 Bq (160×10^{12} Bq)!

The radioactive decay graph of iridium-192 is shown in Figure 15.9.

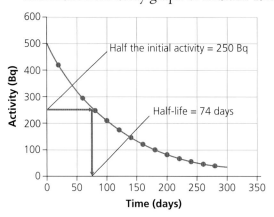

Figure 15.9 Radioactive decay graph of iridium-192

You can see from the graph that the initial activity of the sample of iridium-192 is 500 Bq. The half-life is the time it takes for the activity to go down to (500/2 =) 250 Bq and this happens in 74 days. All radioactive isotopes have decay graph shapes that look like this.

Radiotherapy

In **radiotherapy**, the properties of nuclear radiation are used in the treatment of cancers. There are two types of radiotherapy:

Figure 15.10 External radiotherapy

▶ **External radiotherapy** uses a machine to target an external source of X-rays or gamma rays at the tumour (Figure 15.10). Gamma or X-rays are used because they need to pass from the machine, through the air and penetrate through the skin. The radioisotopes used to produce the radiation have long half-lives so they remain radioactive for a long time.
▶ **Internal radiotherapy** uses radioisotopes with short half-lives. The radioisotopes are chemically attached to drug molecules that target organs in the body. These are injected into the body or swallowed as a drink. Beta emitters are used for internal radiotherapy as they have relatively short penetration into body tissue, killing tumour cells but not the surrounding tissue.

Chemotherapy

Chemotherapy involves using chemical drugs to kill cells in cancerous tumours. The drugs damage cancer cells so they cannot reproduce, reducing the spread of the tumour. Some chemotherapy is non-specific and can kill cancer cells anywhere in the body, and some is targeted at specific organs. Chemotherapy often has serious side effects, making patients feel sick, causing hair-loss and killing healthy cells. Cancer treatments often involve the use of chemotherapy and radiotherapy together.

✔ Test yourself

9 What are the three types of radioactive decay?
10 What can ionising radiation do to living cells?
11 What is the 'half-life' of a radioactive isotope?
12 Table 15.1 shows the radioactive decay of a sample of iodine-131, a radioactive isotope sometimes used to treat thyroid gland problems.

Table 15.1

Time (days)	0	4	8	12	16	20	24	28	32
Activity (Bq)	800	566	400	283	200	141	100	71	50

a) Plot a graph of activity (y-axis) against time (x-axis).
b) Draw a best-fit line (curve) through your points.
c) Use your graph to measure the half-life of iodine-131.

Determination of the half-life of a model radioactive source

A student performed an experiment to determine the half-life of cubes as a model radioactive source. She used 50 cubes, each having one face shaded. A cube has 'decayed' if the shaded face falls facing upwards.

Procedure

1 50 cubes were placed into a plastic tub and then thrown into a plastic tray.
2 All the cubes landing with their shaded face up were considered to have decayed and were removed, counted and recorded.
3 The remaining undecayed cubes were put back into the plastic tub and re-thrown.
4 Step 2 was repeated.
5 Steps 3 and 4 were repeated another 8 times, so that there were 10 throws in total.
6 The results of 10 different students performing the same experiment were collated.

Results

The student collected the following results (Table 15.2):

Table 15.2

Throw	0	1	2	3	4	5	6	7	8	9	10
Number decayed	0	8	7	6	5	4	3	3	2	2	2
Number undecayed	50	42	35	29	24	20	17	14	12	10	9

The collated results from 10 students for the number of undecayed cubes are shown in Table 15.3:

Table 15.3

Throw number	Student										
	1	2	3	4	5	6	7	8	9	10	Total
0	50	50	50	50	50	50	50	50	50	50	
1	42	43	42	42	43	42	42	42	43	42	
2	35	36	36	36	36	36	35	36	36	36	
3	30	30	30	31	30	31	30	30	30	31	
4	25	26	25	26	26	26	26	25	26	26	
5	21	22	22	22	21	22	21	22	22	22	
6	18	18	19	18	19	18	18	18	19	19	
7	15	16	16	16	15	16	15	16	16	15	
8	13	13	14	13	13	13	13	14	13	13	
9	10	11	12	11	11	12	11	11	12	11	
10	9	9	10	10	9	9	10	9	10	9	

Analysing the results

1 Calculate the TOTAL number of undecayed cubes for each throw.
2 Plot a graph of total undecayed cubes (y-axis) against throw number (x-axis).
3 Draw a suitable best-fit line.
4 Use the graph to determine the half-life of the cubes.

⬇ Chapter summary

- Gamma rays and X-rays from the electromagnetic spectrum are used in the diagnosis of disease and injury.
- Drug treatments can have positive effects and possible negative side effects.
- New drugs undergo clinical testing before they are released for use on the general population. The clinical trial process can involve testing on animals, which raises ethical issues.
- Radioactive emissions from radioisotopes and the short wavelength parts of the electromagnetic spectrum are called ionising radiation. They harm or kill living cells.
- Gamma cameras can detect gamma rays emitted by a radioisotope source from cancer cells. The radioisotopes can be carried to target organs of the body by drugs that only affect the target organ.
- Radioisotopes can emit: alpha particles (helium nuclei); beta particles (high velocity electrons ejected from the nucleus of a decaying atom); or gamma radiation. These have different penetrating and ionising powers.

- Ionising radiation can interact with atoms or molecules in living cells, causing damage to DNA. Cancer cells are more susceptible to damage and die or reproduce more slowly – so ionising radiation can be used to treat cancers; this is called radiotherapy.
- External radiotherapy uses an external source of X-rays or gamma rays to target and treat a tumour. Internal radiotherapy uses a radioisotope, which can be taken as a drink or injected into a vein.
- The half-life of a radioisotope is the time taken for the activity of a sample of the radioisotope to halve.
- The emissions and half-life of a radioisotope can be used to select the most suitable radioisotope for a medical purpose.
- Chemotherapy uses drugs to kill cancer cells and is often used together with radiotherapy.
- Medical imaging uses electromagnetic radiation or sound waves to create images of the human body to reveal, diagnose or examine disease.
- Ultrasound, X-rays (both standard X-ray examinations and CAT scanners) and MRI are all methods of medical imaging.

16 Fighting disease

As we have learnt to treat infection and diseases, life expectancy has risen. However, new diseases arise, and new treatments and vaccinations can be developed. Knowing how our bodies resist infections helps in this development.

▶ Microorganisms and disease

The term microorganism is used to describe organisms that can only be seen with a microscope, including viruses, bacteria, microscopic fungi, and protists. There are about 10 million trillion microorganisms for every human on the planet; some are 'good' for human health, some are 'bad' and most are neither good nor bad.

Some microorganisms can cause disease and some cause inconvenience, like spoiling our food. Those that cause disease are called pathogens. Many microorganisms perform vital functions. We have bacteria on our skin that help to keep it in good condition, and others in our gut that help with digestion. The spoiling of food is just an unfortunate side effect of a vital function that microorganisms perform, which is breaking down dead organisms so that the nutrients in them can be recycled.

Defences against infection

The human body defends itself against pathogens in two ways. First, there are features that stop pathogens entering the body. If that is unsuccessful, the immune system kicks in to kill any pathogens that enter.

Human skin is an impenetrable barrier, covering nearly all the body. When intact, it works very well at stopping the entry of microorganisms. If the skin is damaged, blood clots to plug the gap, sealing off the wound while the skin heals and restores the barrier. However, blood clotting is not quick enough to stop the entry of microorganisms completely. Pathogens can also enter through the body openings, which have no skin.

Communities of microorganisms, called the skin flora, live on the surface of our skin. They are well adapted to the habitat, so it can be difficult for any pathogen to establish themselves.

Once pathogens get into the body, white blood cells work to kill them. The blood is an ideal place for our immune system, as it circulates to all parts of the body and can reach any infections. Two types of white blood cells (Figure 16.1) attack invading microorganisms:

▶ Phagocytes ingest ('eat') microorganisms and digest them.
▶ Lymphocytes produce chemicals called antibodies, which destroy microorganisms, and antitoxins, which neutralise any

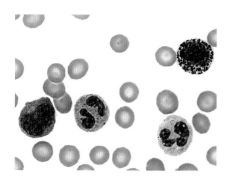

Figure 16.1 A blood smear showing a phagocyte and a lymphocyte

poisons produced by the pathogens. (One reason why pathogens cause disease is because they produce waste chemicals that are poisonous to human cells.)

▶ Antigens and antibodies

To attack invading microorganisms, the immune system must identify them and distinguish them from the body's own cells. All cells have patterns of molecules called antigens on their surface, and the pattern is different for every individual. If white blood cells come across invading cells that do not have the usual pattern of antigens, they attack them.

Phagocytes attack any cells with foreign antigens. Lymphocytes respond by producing specific chemicals called antibodies; the antibodies they produce depend on the antigens detected. The antibodies can destroy the microorganisms or stick them together so that phagocytes can ingest a lot at once.

Memory cells

(H) When the body encounters a new pathogen, there are no specific antibodies for its antigens. The phagocytes attack, but the lymphocytes take a while to develop antibodies. During this time, the pathogen may reach levels that cause symptoms of disease. Once the lymphocytes have made the antibodies, however, they form memory cells for that disease. If the same pathogen is encountered again, the memory cells quickly produce the appropriate antibodies, and the pathogen is wiped out before the infection takes hold.

Once antibodies are produced, the person becomes immune to the disease. Sometimes, as in the case of measles, that immunity lasts a lifetime. Some microorganisms, however, can get around this immunity if a **mutation** occurs so they produce different antigens. For example, the flu virus mutates often and rapidly, producing new antigens. You can have flu multiple times, as your body fails to identify these new antigens that are different from previous infections.

> ### ✔ Test yourself
>
> 1 What is meant by the term 'pathogen'?
> 2 Explain the role of blood clotting in defending the body.
> 3 Which type of white blood cell produces antibodies.
> 4 Explain why people may get flu many times.

▶ Vaccination

Although we can naturally become immune to diseases, it is better not to catch some serious infections at all. We can get immunity to both bacterial and viral diseases by having vaccinations.

When you are vaccinated against a disease, you are injected with dead or weakened pathogens that are incapable of causing symptoms. These still have antigens, so your lymphocytes can react and build up memory cells, and you become immune to that disease. The inactivated microorganisms cannot reproduce inside the body, so the immune response to the vaccine is not as great as when you get the disease itself. 'Booster' injections are given after the first vaccination to increase the immune response. Sometimes, rather than using the whole pathogen, vaccines contain only antigens or parts of antigens.

Making decisions about vaccines

Many people think it makes sense for children to have immunity to serious diseases and the decision to vaccinate is made by parents. Sometimes this is not an easy decision:

▶ The vaccination often involves an injection, which can scare or hurt young children.
▶ Vaccinations usually have some side effects. These are usually minor, for example, inflammation of the injection site or briefly feeling unwell; rarely, they may be more serious, such as allergic reactions.
▶ In 1998, a scientifically unsound study on the MMR vaccine (against measles, mumps and rubella) proposed a link with autism. Many parents decided not to vaccinate their children and the incidence of measles (a serious, even fatal disease) grew over the next 15 years. More recently vaccination rates have recovered.

When making decisions, people should consider the risks of vaccination against the risks from the disease. It is important to understand that:

▶ Like all medicines and treatments, vaccines must meet stringent safety standards and are clinically tested before being approved for use (page 124).
▶ Vaccines do have side effects but are only approved if these are mild and/or extremely rare.
▶ The best advice comes from doctors and scientists who specialise in disease. Media reports may not be accurate or scientific. Social media and public opinion are not reliable.

▶ Antibiotics

Antibiotics are chemicals which destroy or slow the growth of bacteria. Most antibiotics have been developed from products made by living organisms; penicillin is produced naturally by a fungus. Penicillin for clinical use is synthesised from the natural product. It works by weakening bacterial cell walls so they take in water and burst (other antibiotics work in different ways). When a patient takes antibiotics, they travel around in the blood. They are ineffective against viruses because viruses enter the body's cells so antibiotics cannot reach them.

Antibiotic resistance

Some species of bacteria have evolved resistance to most of the antibiotics in clinical use. The media call these bacteria 'superbugs'. MRSA (methicillin-resistant *Staphylococcus aureus*; Figure 16.2) is a resistant strain of a common bacterium, usually carried on the skin, where it can cause boils and mild skin infections. If it gets

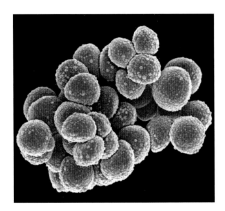

Figure 16.2 Methicillin-resistant *Staphylococcus aureus* bacteria, MRSA

through the skin it can cause life-threatening conditions such as blood poisoning.

A mutation can give bacteria resistance to an antibiotic. Bacteria reproduce very quickly, so a new resistant gene caused by mutation can spread rapidly. Widespread use of the antibiotic has the effect of killing all the susceptible bacteria, leaving only the ones that are resistant. The antibiotic becomes useless for fighting that species of bacteria.

MRSA is particularly dangerous in hospitals, where people are already unwell or have wounds from accidents or surgery. The control measures for MRSA fall into two categories – the prevention of infection and combating the evolution of resistance.

To prevent infection:

▶ Patients entering hospital are screened for MRSA.
▶ Hospital staff wash their hands after going to the toilet, before and after eating and before touching patients. Good hygiene should also be practised by the general public.
▶ Visitors are recommended to wash their hands or use hand sanitiser gels when entering wards.
▶ Stringent hygiene measures are used with any procedure involving body openings or wounds.

To slow the development of resistance:

▶ Doctors avoid prescribing antibiotics wherever possible – for example, if an infection is mild and the body can overcome it without antibiotics.
▶ When antibiotics are prescribed, doctors vary the type. Extensive use of any single antibiotic increases the risk that it will become ineffective.

✔ Test yourself

5 Explain why a vaccine gives immunity, even when the pathogen it contains is dead.
6 Explain why 'booster' vaccinations are often necessary for full immunity.

7 What is the difference between an antibody and an antibiotic?
8 Why can a bacterial population become immune to an antibiotic so quickly?

⚙ Specified practical

Investigation into the effects of antibiotics on bacterial growth

Students were provided with a pre-prepared agar plate seeded with bacteria, four different antimicrobial agents, labelled A–D, and four sterile paper discs. They carried out the following procedure, after first washing their hands and wiping down the working area with disinfectant.

Procedure

1 The students worked very close to a lit Bunsen burner. Forceps were flamed and used to pick up a filter paper disc and dip it into antibiotic A.

2 The disc was allowed to dry for 5 minutes on an open, sterile Petri dish, next to a lit Bunsen burner.
3 Step 3 was repeated for antibiotics B, C and D.
4 An agar plate, already seeded with bacteria, was held upside down. The base was divided into four sections, labelled A, B, C, D, by drawing a cross with the marker pen (Figure 16.3).

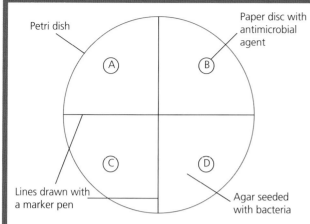

Petri dish — A

Paper disc with antimicrobial agent — B

Lines drawn with a marker pen — C

Agar seeded with bacteria — D

Figure 16.3 Prepared Petri dish

5 Forceps were flamed and then used to pick up antibiotic disc A. With the Petri dish the right way up, the lid was raised at an angle and the disc was placed onto the agar in the centre of section A.

6 Step 5 was repeated for the other three discs.

7 The lid was taped securely and the dishes were incubated (inverted) for 2–3 days at 20–25 °C.

8 The plates were observed without opening them and the width of the clear zone around each antimicrobial agent was recorded.

Analysing the results

1 Why is it recommended to work very close to a lit Bunsen burner?

2 Explain the reason for raising the Petri dish lid at an angle.

3 Explain the reason why the incubation temperature should be kept well below 37 °C.

4 Explain in detail why more effective antimicrobial agents produce larger clear zones around the disc.

Chapter summary

- Some microorganisms are harmless and perform vital functions, but some, called pathogens, cause diseases.
- Intact skin is a barrier against microorganisms. Blood clots seal wounds and prevent microorganisms entering the body.
- Some white cells in the blood ingest microorganisms **H** and others produce antibodies and antitoxins.
- Pathogens may have to compete with the body's natural population of microorganisms.
- Vaccination protects people from infectious diseases.
- Certain factors may influence parents' decisions about whether to have children vaccinated or not. There is a need for sound scientific evidence when making these decisions, rather than relying on the media and public opinion.
- Antigens are molecules that are recognised by the immune system. Some white blood cells, lymphocytes, recognise antigens and secrete antigen-specific antibodies.

- Antibodies destroy pathogens or aid white blood cells to destroy them.
- Vaccines contain inactivated pathogens or antigens or parts of antigens that are derived from disease-causing organisms. Vaccines stimulate antibody production to protect against bacteria and viruses.
- Memory cells that are produced following natural infection or vaccination can make specific antibodies very quickly if the same antigen is encountered a second time.
- Flu can occur multiple times, but immunity to measles is usually lifelong.
- Antibiotics, including penicillin, are chemicals originally produced by living organisms such as fungi. Antibiotics kill infecting bacteria or prevent their growth.
- Resistance in bacteria, such as in MRSA, results from overuse of antibiotics. Doctors and hospitals use control measures for MRSA.

Exercise and fitness in humans

Exercise is good for our health. When humans move, we do work that requires energy. Movement is caused by muscle contraction, which is controlled by the nervous system.

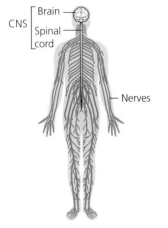

Figure 17.1 The human nervous system

▶ The nervous system

The nervous system consists of two parts (Figure 17.1):

- ▶ The **central nervous system (CNS)**: the brain and the spinal cord.
- ▶ The **peripheral nervous system**: all the nerves which branch off the central nervous system and spread throughout the body, including to muscles.

Muscles are controlled by nerve impulses, which are electrical signals carried by the **neurones** (nerve cells).

Nerve impulses also pass back from the muscles to the CNS, to give the brain information about the muscles.

▶ The skeleton and movement

The human skeleton consists of bone and cartilage. Its functions are protection of vital organs, providing attachment points for muscles, and support for the body by means of rigid bones. Separate bones with moveable joints allow for movement. There are different types of joints in the skeleton:

- ▶ **Fixed joints** between the bones in the skull do not move.
- ▶ **Hinge joints** in the elbow and knee work like hinges and allow movement in one direction.
- ▶ **Ball and socket joints** in the shoulder and hips allow bones to rotate.

The structure of a synovial joint

Many of the moveable joints in the body are synovial joints, lubricated by synovial fluid. The structure of a synovial joint and the functions of its parts are shown in Figure 17.2.

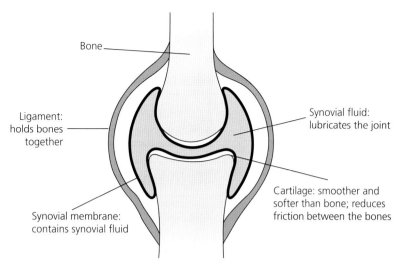

Bone

Ligament: holds bones together

Synovial fluid: lubricates the joint

Cartilage: smoother and softer than bone; reduces friction between the bones

Synovial membrane: contains synovial fluid

Figure 17.2 A synovial joint

Antagonistic muscles

Muscles only do work when they contract – they can pull but never push. This means that one muscle moves a joint in one direction and another one moves it back. These pairs of opposing muscles, for example the muscles that move the forearm, are called antagonistic pairs. The biceps muscle contracts to bend the elbow, and the antagonistic triceps muscle contracts to straighten it again (Figure 17.3).

Health issue with bones and joints

Joints can be damaged by disease and injury. A common joint disease is osteoarthritis, where the cartilage on the ends of bones breaks down. The bones grind together causing pain, swelling and movement problems.

Several factors increase the risk of developing osteoarthritis: age (risk increases as you get older), previous joint injuries, family history, gender (it is more common in women) and obesity.

Injuries to joints, particularly common in sportspeople, include torn ligaments and damaged cartilages. Damaged cartilage causes temporary symptoms similar to osteoarthritis but the injury can respond to rest. Severe cartilage damage requires surgery. Ligament damage can vary from a stretched ligament (a sprain) to partial or complete tears. As with cartilage injuries, minor ligament injuries heal on their own, but more severe injuries require surgery.

Badly damaged joints can be surgically replaced with artificial joints made of plastic and metal.

There are three main types of broken bone (Figure 17.4):

▶ **Simple fracture** – the bone may be cracked or completely broken, but the skin is not pierced by the broken bone.
▶ **Compound fracture** – the broken bone pierces the skin.
▶ **Greenstick fracture** – the bone bends and breaks on one side only. These fractures are most common in young children, as their bones are softer.

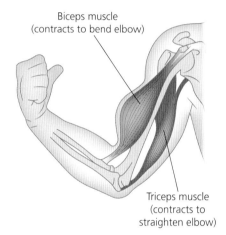

Biceps muscle (contracts to bend elbow)

Triceps muscle (contracts to straighten elbow)

Figure 17.3 When a muscle contracts, its antagonistic muscle relaxes

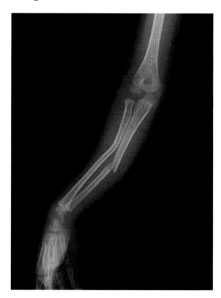

Figure 17.4 Simple fracture

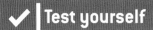
Mathematical information about movement

The motion of an object can be described using the following quantities:

- **distance** (measured in metres, m): how far the object travels or how far away the object is from a fixed point
- **time** (measured in seconds, s): the time interval between two events or the time since the start of the motion
- **speed** (measured in metres per second, m/s): a measure of how fast or slow the object is moving. The speed of the object can be calculated using the equation:

$$\text{speed} = \frac{\text{distance}}{\text{time}}$$

- **velocity** (measured in metres per second, m/s, in a given direction): a measure of how fast or slow (the speed) the object is moving in a given direction (left, right, north or south)
- **acceleration** or deceleration (measured in metres per second per second, m/s^2): the rate that the object is speeding up or slowing down, which is the rate of change of velocity. Acceleration can be calculated using the equation:

$$\text{acceleration or deceleration} = \frac{\text{change in velocity}}{\text{time}}$$

Speed is a **scalar** quantity because it only has magnitude (size); velocity is a **vector** quantity because it has direction as well as magnitude.

★ Worked examples

1 Usain Bolt holds the men's 200 m world record, running it in 19.19 seconds in 2009. What was his average speed?
2 In a 100 m race, Usain Bolt reached a velocity of 28.5 m/s after 3 seconds. What was his acceleration?

Graphs of motion

The motion of objects can be described and analysed using graphs. There are two types of motion graph: distance–time graphs and velocity–time graphs.

Distance–time graphs

A distance–time graph allows us to measure the speed of a moving object. Figure 17.5 shows the distance–time graph of a person running at a constant speed of 3 m/s.

Stationary objects are represented by straight horizontal lines. The slope or gradient of a distance–time graph is the speed of the object.

Velocity–time graphs

A velocity–time graph gives us more information than a distance–time graph. The graph in Figure 17.6 shows a cyclist that is:

▷ stationary for 2 seconds
▷ accelerating at 3 m/s² for 2 seconds
▷ moving at a constant velocity of 6 m/s for 6 s

The slope or gradient of a velocity–time graph is the acceleration of the object. The distance travelled by the object is the area under the velocity–time graph (in Figure 17.6, this is 42 m).

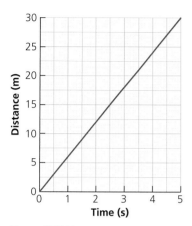

Figure 17.5 Distance–time graph of a runner

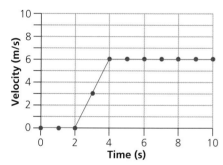

Figure 17.6 Velocity–time graph of a cyclist

⚙️ Specified practical

Determination of the acceleration of a moving object

Students studied the acceleration of a squash ball travelling down a ramp. The apparatus is shown in Figure 17.7.

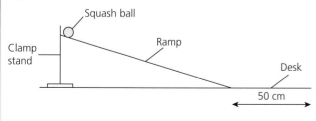

Figure 17.7

Procedure

1 The ramp was set at a height of 10 cm above the desk.

2 A mark was made 50 cm from the end of the ramp.

3 The squash ball was released from the top of the ramp and the stopwatch was started at the same time.

4 The lap button on the stopwatch was pressed when the ball reached the bottom of the ramp.

5 The stopwatch was stopped when the squash ball reached the 50 cm mark.

6 The time taken for the ball to travel down the ramp (lap time) and the total time were recorded.

7 Steps 1–6 were repeated, increasing the height in 5 cm intervals each time up to 25 cm.

8 The whole experiment was repeated twice more.

Results

Students were then asked to process the results in the following way.

1 Calculate the time taken for the ball to travel 50 cm along the bench; this is the total time – the lap time.
2 Calculate the velocity at the bottom of the ramp using the formula

velocity

$$= \frac{0.5}{\text{mean time taken to travel 50 cm along the bench}}$$

3 acceleration

$$= \frac{\text{velocity at bottom of ramp}}{\text{mean time to reach the bottom of the ramp}}$$

4 Plot a graph of ramp height against acceleration.

Analysing the results

1 Explain the reason why the time taken for the ball to move 50 cm after leaving the ramp was measured.
2 Suggest possible sources of error in this experiment (excluding human error).
3 The measurements could be taken using light gates and a data logger. Suggest why this might improve the quality of the results.
4 What should be compared to assess:
   a) The repeatability of this experiment?
   b) The reproducibility of this experiment?

Test yourself

5 Explain the difference between speed and velocity.
6 A horse walks at 2 m/s and trots at 3.8 m/s. It travels in a straight line, walking for 30 seconds and then trotting for 30 seconds. How far will it travel?
7 On a velocity–time graph of a person walking, what does a straight, horizontal line tell you about the movement of the person?

The cardiovascular system

The cardiovascular system consists of the heart and blood vessels (arteries, veins and capillaries). It is the body's transport system, carrying oxygen, carbon dioxide, food chemicals and wastes from one part of the body to another. Because the blood system reaches every part of the body it also plays a useful role in the immune system (page 129).

Blood pumped by the heart moves around the body in arteries, veins and capillaries. When it leaves the heart, the blood travels to the organs in arteries. The arteries branch into many small capillaries, which take the blood through the organs. The capillaries join to form veins, which carry the blood back to the heart.

The structure of the circulatory system of a mammal is shown in Figure 17.8. Note that the diagram has been simplified.

In mammals the heart is divided into two halves. The left half receives blood from the lungs and pumps it to the rest of the body. The right half receives blood from the body and pumps it to the lungs. The blood travels in two separate circuits (a double circulatory system) around the body: the pulmonary circulation to and from the lungs and the systemic circulation to and from the rest of the body (Figure 17.8).

Key term

Double circulatory system Blood system in which the blood travels through the heart twice on each circuit of the body.

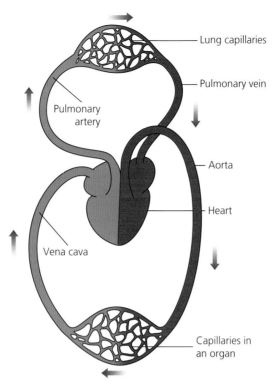

Figure 17.8 Structure of the circulatory system of a mammal. Arrows show the direction of blood flow

The heart

Blood is moved around the body by the pumping of the heart. When the heart muscle contracts, it applies a force and pushes the blood out into the arteries.

Figure 17.9 shows how the blood flows through the heart. The right side pumps de-oxygenated blood and the left side pumps oxygenated blood. The left ventricle has a much thicker wall than the right ventricle, because it must pump blood all around the body, whereas the right ventricle only has to pump blood to the lungs.

Blood flows through the heart in one direction only, from the atria to the ventricles and then out of the arteries at the top. The valves between the atria and the ventricles stop backflow from the ventricles into the atria, and the valves at the beginning of the aorta and the pulmonary artery make sure that blood that has left the heart is not sucked back when the ventricles relax.

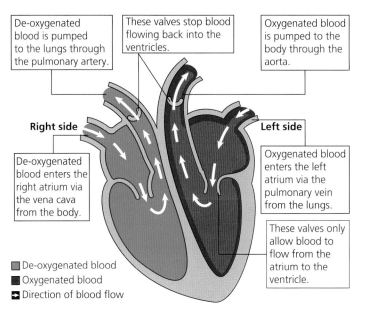

De-oxygenated blood is pumped to the lungs through the pulmonary artery.

These valves stop blood flowing back into the ventricles.

Oxygenated blood is pumped to the body through the aorta.

Right side

De-oxygenated blood enters the right atrium via the vena cava from the body.

Left side

Oxygenated blood enters the left atrium via the pulmonary vein from the lungs.

These valves only allow blood to flow from the atrium to the ventricle.

☐ De-oxygenated blood
■ Oxygenated blood
➡ Direction of blood flow

Figure 17.9 Blood flow through the heart

The blood vessels

There are three types of blood vessels – arteries, veins and capillaries. Each has a different structure, which is linked to its functions.

Arteries carry blood at high pressure away from the heart. The arteries must be able to resist that pressure. Arteries deliver blood to a large number of small capillaries, which take blood through the organs. Exchange of materials occurs in the capillaries. Oxygen and nutrients are delivered to cells and waste products (including carbon dioxide) are picked up. Capillaries are very narrow, so blood flows slowly and materials can be exchanged.

The capillaries discharge blood into veins, which take it back to the heart. In veins there is no pulse and the blood pressure has dropped. Blood is returned to the heart at the same rate as it leaves, yet veins have no pulse. Blood is moved in the veins by the body's muscles. The contraction of these muscles, while performing their normal functions, squeezes the thin-walled veins and so moves the blood. The direction of blood flow is controlled by the valves, which prevent the flow of blood back towards the capillaries. Arteries do not need valves because the pulse ensures that the blood flows in the right direction.

Table 17.1 shows the features of the different blood vessels, and how their structure links to their functions. The structure of the vessels is also shown in Figure 17.10.

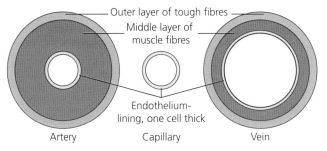

Outer layer of tough fibres
Middle layer of muscle fibres
Endothelium-lining, one cell thick

Artery   Capillary   Vein

Figure 17.10 Structure of an artery, a capillary and a vein (drawings are not to scale)

Table 17.1 Structure and function of blood vessels

Vessel	Structural feature	Link to function
Artery	Thick muscular wall	Resists high blood pressure
	Pulse	Pushes blood through the vessel
Vein	Thinner wall than an artery	Does not need to resist high blood pressure – allows muscles around the vessel to squeeze the blood and cause it to move
	Valves	Ensures that movement of blood is only towards the heart
	Large lumen (gap in the middle of the vessel)	Increases the rate of flow of the blood
Capillary	Wall is only one cell thick	Allows easy passage of materials in and out by diffusion
	Blood flow is very slow	Allows time for materials to be exchanged
	Extensive networks in each organ	Every cell is near to a capillary – more materials can be exchanged

Blood

The components of blood are:

▷ **Plasma** – the liquid part of the blood, which transports water-soluble substances including digested food, carbon dioxide, urea, salts and hormones.
▷ **Red blood cells** – these cells carry oxygen around the body, attached to the red pigment haemoglobin.
▷ **White blood cells** – there are various types of white blood cells, including phagocytes and lymphocytes (page 129).
▷ **Platelets** – these are cell fragments that help the blood to clot. Clotting helps prevent infection (page 129).

Red blood cells are biconcave discs – round and flattened, with a central indentation (Figure 17.11). Their shape increases the surface area for absorption of oxygen. They are red because they contain the haemoglobin, which absorbs oxygen. Mature red blood cells are unusual because they have no nucleus, allowing more haemoglobin to be packed into their cytoplasm.

Phagocytes ingest bacteria (Figure 17.12). These white blood cells can change shape and can also move, squeezing through tiny gaps in capillary walls and entering tissue fluid to fight infections.

> **Key term**
>
> **Haemoglobin** Red pigment in red blood cells that carries oxygen.

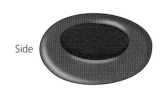
Side

Surface

Figure 17.11 Structure of a red blood cell

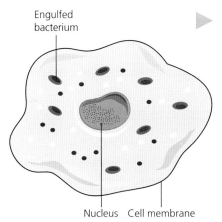
Engulfed bacterium

Nucleus   Cell membrane

Figure 17.12 Structure of a phagocyte (a type of white blood cell)

Measuring the body's response to exercise

Physical fitness can be measured by collecting data before, during and after exercise:

▷ **Pulse rate** can be used as a measure of heart rate. Normal resting heart rate for an adult is between 60–100 beats per minute (bpm) but varies with age, gender, and physical fitness.
▷ **Recovery time** is the time taken for the heart rate to return to its resting value after exercise. Increased physical fitness decreases recovery time.
▷ **Breathing rate** increases during exercise to provide the extra oxygen needed by muscles. During intense exercise, breathing rate may increase from a typical resting rate of 15 breaths per minute up to 40–50 breaths per minute.

Over the long term, training increases the efficiency of the use of oxygen by the muscles and the heart gets stronger, meaning that it can deliver more blood per beat to the muscles. The breathing rate and recovery time decrease, as does the resting pulse rate.

✔ **Test yourself**

8 In the heart, why is the wall of the left ventricle thicker than that of the right?

9 Explain why valves are necessary in veins but not in arteries.

10 Which components of the blood are involved in transporting substances?

11 Suggest a reason why recovery time is a better measure of fitness than pulse rate.

⬇ **Chapter summary**

- Energy is needed by muscles to do work.
- The nervous system is composed of the central nervous system (the brain and spinal cord) and the peripheral nervous system.
- Nerve impulses (electrical signals carried by nerve cells, or neurones) are the cause of muscle contraction.
- Antagonistic muscles (such as biceps and triceps) move joints in opposite directions.
- A synovial joint has cartilage, ligaments, synovial fluid and synovial membrane.
- Joint disease (osteoarthritis) and injury (torn ligaments) can result in limited movement of joints. Badly damaged joints can be replaced by artificial joints.
- There are different types of bone fracture: simple, compound and greenstick.
- Fixed joints are found in the skull, hinge joints in the elbow and knee, and ball and socket joints in the shoulder and hip.
- Distance–time and velocity–time graphs can be used to analyse movement.
- Mathematical equations can provide information relating to movement:

$$\text{speed} = \frac{\text{distance}}{\text{time}}$$

$$\text{acceleration or deceleration} = \frac{\text{change in velocity}}{\text{time}}$$

- Velocity–time graphs can be used to determine acceleration and distance travelled.
- The human cardiovascular system includes the heart (with ventricles, valves and atria), veins, arteries, capillaries.
- The human blood system is a double circulatory system.
- The structure of blood vessels is related to function (arteries have thick, muscular walls; veins have thinner walls and valves to prevent backflow of blood; capillaries are one cell thick to allow exchange of substances).
- Blood is composed of red blood cells (containing haemoglobin to transport oxygen), white blood cells to fight infections, plasma to transport substances and platelets which play a role in clotting.
- Measurements can be taken to monitor pulse rate, breathing rate and recovery time.
- Exercise affects breathing (short term effects: breathing rate increases to provide the oxygen and remove carbon dioxide; long term effects: the body becomes more efficient at transporting oxygen).
- There are physiological effects of exercise on heart rate and recovery time (short term effects: heart rate increases, cardiac output increases; long term effects: heart muscle is strengthened, heart muscle becomes more efficient).

▶ Practice exam questions

1 Gregor Mendel worked with pea plants. In peas, a green pod is dominant to a yellow pod. Mendel crossed two heterozygous pea plants (containing one allele for green and one for yellow), both with green pods, but each carrying what he called a yellow 'factor'. He repeated this cross 580 times, and the plants produced 428 green pods and 152 yellow pods.

   a) What do we call Mendel's 'factors' today? [1]

   b) Represent the green 'factor' with G and yellow with g. Draw a Punnett square of the cross that Mendel did (Gg × Gg) and state the expected ratio of green pods to yellow pods. [3]

   c) What was the actual ratio of green pods : yellow pods in Mendel's results? [2]

   d) The actual ratio does not match the expected ratio. Why does this not mean that the expected ratio is wrong? [1]

   e) Of the 428 green pods, how many (approximately) would you expect to be homozygous (GG)? [2]

2 The table below gives the energy requirement of human males at different ages:

Age (years)	Energy requirement per day (kcal/day)
5	1482
10	2032
15	2820
19–24	2772
45–54	2581
65–74	2294

   a) Suggest a reason why the energy requirement goes down after age 15? [2]

   b) The mean weight of males between 19–24 is 76 kg. Calculate their energy requirement per kg of body weight. [2]

   c) The mean energy requirement of 15-year-old girls is 2390 kcal/day. Suggest why this figure is lower than for boys. [2]

   d) When comparing the energy requirements of 5-year-olds and adults, suggest a reason why it would be better to use kilocalories per kg of body weight per day, rather than just kcal/day. [1]

3 8 students were modelling radioactive decay using cubes. Each student was given 50 cubes and each cube had a white spot on one face. They used the following method:

   a) Each student threw 50 cubes.

   b) All cubes with a white spot facing upwards were removed.

   c) The number of remaining cubes was counted and recorded.

   d) The remaining cubes were thrown.

   e) Steps b), c) and d) were repeated until the cubes had been thrown 8 times.

The table shows one student's results; and the combined results of all the students.

Throw number	Number of cubes remaining	
	One student's results	Combined results
0	50	400
1	42	340
2	36	290
3	30	240
4	26	210
5	22	170
6	18	150
7	16	130
8	13	110

   a) David said that the combined set of results were more repeatable than a single student's set of results. State if you agree or disagree with David. Explain your answer. [2]

   b) Part of the combined results have been plotted. Re-plot the graph with the remaining data and draw a suitable line. [3]

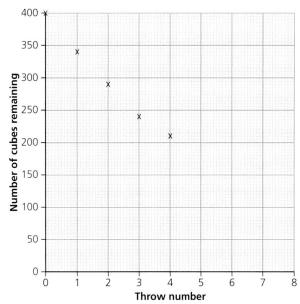

   c) David used the completed graph from b) to determine the half-life of the cubes. Use the completed graph to determine the half-life. Show your working on the graph. [2]

d) Protactinium-234 has a half-life of 77 s. The initial activity of a sample was measured to be 80 counts per second. Calculate the time for the activity to fall to 10 counts per second. [2]

4 The decay method and half-life of some radioisotopes used for medical purposes are shown in the table.

Radioisotope	Decay method	Half-life
Carbon-14	Beta	5730 years
Tellurium-133	Beta	12 minutes
Technecium-99	Gamma	6 hours
Cobalt-60	Beta and gamma	5 years
Americium-241	Alpha	432 years
Astatine-211	Alpha	7.2 hours

Select the most suitable radioisotope for the tasks below and give reasons for your choice. Only use data from the table.

a) Internal radiotherapy on a cancer tumour by injecting the radioisotope directly into the tumour. [3]

b) A source of gamma rays for external radiotherapy. [3]

c) A radiotracer for detecting blood flow through a kidney. [3]

5 Cattle were given a vaccine against ringworm, a fungal infection. The levels of antibodies in their blood were monitored over a period which included a booster vaccine. The results are shown below.

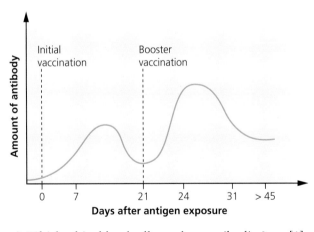

a) Which white blood cells produce antibodies? [1]

b) What term would be used to describe ringworm, as an organism which can cause disease? [1]

c) How do the blood cells recognise ringworm, so that antibody production is triggered? [1]

d) What evidence is there that the cows will need a further booster vaccine? [1]

e) Cattle are sometimes given antibiotics as a precaution, even though they are healthy. Explain why this practice is being discouraged in an effort to slow down the evolution of 'superbugs'. [3]

6 The diagram shows the bones and muscles in the upper arm.

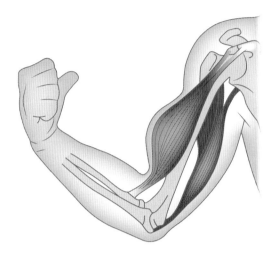

a) Name the **two types** of joint shown in the diagram. [2]

b) State the conditions (contracted or relaxed) of the biceps and triceps muscles when the arm is straightened. [1]

c) The biceps and triceps are antagonistic muscles. State the meaning of this term. [1]

d) Name the following components of a joint like the shoulder joint:

i) The liquid that lubricates the joint. [1]

ii) The tissue that holds the bones of the joint together. [1]

iii) The tissue that reduces friction in the joint. [1]

7 Varicose veins are caused by a weakening and stretching of the walls of the veins and the valves become less effective. The veins of the legs become swollen and enlarged, appearing blue or dark purple.

a) What is the function of the valves in veins? [1]

b) Why is there no need for valves in arteries? [1]

c) Suggest why varicose veins are worse after a long period of standing. [2]

d) Suggest why the legs of a person with varicose veins feel heavy. [1]

e) Exercise is one aspect of the treatment of varicose veins. During exercise the muscles are constantly contracting. Suggest a reason why this might help with varicose veins. [1]

18 Materials for a purpose

When designing objects, designers have to take into account the properties of the materials that they intend to use. Using the wrong material can seriously affect the look, feel and functionality of objects.

▶ Bonding

On an atomic and molecular level, materials consist of particles (atoms and molecules), bonded together. The main types of bonding are ionic, covalent and metallic.

Ionic bonding

Ionic compounds are ones where the particles are joined by ionic bonds. Ionic bonds are formed by the electrostatic attraction between oppositely charged particles (**ions**). Atoms are most stable when they have a full outer shell of electrons. Atoms with relatively full outer shells (non-metals) tend to gain electrons and so acquire a negative charge (becoming **anions**). Atoms with very few electrons in their outer shell (metals) tend to lose those electrons and become positively charged (or **cations**). If metallic and non-metallic atoms react together, the metal will donate one or more electron to the non-metal, and the two charged ions will then bond together. Figure 18.1 (called a 'dot and cross' diagram) illustrates this process for the formation of sodium chloride.

Key terms

Ionic compound Formed between particles joined by ionic bonds.
Electrostatic attraction Ionic bonds are formed by this attraction between oppositely charged particles (ions).

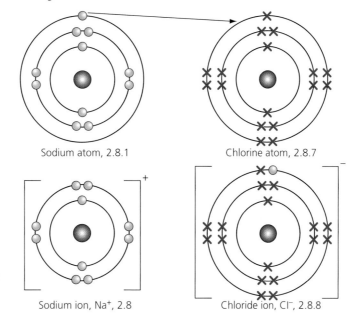

Sodium atom, 2.8.1

Chlorine atom, 2.8.7

Sodium ion, Na⁺, 2.8

Chloride ion, Cl⁻, 2.8.8

Figure 18.1 When sodium and chlorine atoms react, the sodium atom loses an electron forming a positively charged sodium ion, Na^+ (a cation) and the chlorine atom gains an electron forming a negatively charged chlorine ion, Cl^- (an anion)

Both ions now have full outer shells and there will be a strong electrostatic attraction between the oppositely charged ions, which is called an ionic bond. Ionic bonds are formed by the transfer of electrons from one atom to another.

The charge on an ion of an element depends on which Group of the Periodic Table it is found in. Group 1, 2 and 3 elements lose 1, 2 or 3 electrons respectively, (forming +1, +2 or +3 charges), and Groups 5, 6 and 7 gain 3, 2 and 1 electron(s) respectively, (forming −3, −2 and −1 charges). Elements in Group 4 tend not to form ions, and elements in Group 0 are stable and do not lose or gain electrons.

Properties of ionic compounds

Sodium chloride is an example of a typical ionic compound. The millions of sodium and chloride ions in a crystal of sodium chloride are held together in a regular three-dimensional lattice (a regular, repeating structure) by strong electrostatic forces (see Figure 18.2). This is referred to as a giant ionic structure.

The high melting points of ionic compounds can be explained by the fact that each anion attracts all the cations around it, and vice versa, so a lot of energy is needed to overcome the electrostatic attractive force. This energy can be provided by heating, but high temperatures are needed to supply enough energy to break the ionic bonds.

Solid ionic compounds do not conduct electricity because the ions are held in fixed positions within their structure and are not free to move and conduct electricity. When molten or dissolved in water, they do conduct electricity because the structure breaks down, and the ions are free to move – and create an electrical current. Many, but not all, ionic compounds are soluble in water.

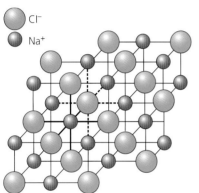

Cl⁻
Na⁺

Figure 18.2 Sodium and chlorine atoms and ions

✔ Test yourself

1. What is an ion?
2. Calcium (Ca) is in Group 2 of the Periodic Table. What is the formula for a calcium ion? Explain your answer.
3. How are ionic bonds formed?
4. Molten sodium chloride and sodium chloride solution both conduct electricity, but solid sodium chloride does not. Explain why.
5. Calcium oxide is an ionic compound. Calcium has two electrons in its outer shell, and oxygen has six electrons in its outer shell. Draw dot and cross diagrams to show how calcium oxide is formed by electron transfer from calcium to oxygen atoms.

Covalent bonding

Atoms that need to lose or gain three or four electrons to have a full outer shell rarely or never form ions. They make **molecules** by forming **covalent bonds**, in which electrons are shared. Water is an example of a covalent molecule.

Water molecules form when hydrogen reacts with oxygen; two atoms of hydrogen combine with one atom of oxygen. Hydrogen has one electron in its outer shell and needs to gain one to fill its

Key term

Covalent bonds Bonds formed between atoms that share electrons.

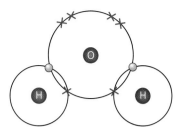

Figure 18.3 The dot and cross diagram for a molecule of water

outer shell. Oxygen has six electrons in its outer shell and needs to gain two to fill its outer shell.

Figure 18.3 shows the electron dot and cross diagram for hydrogen and oxygen forming water. The oxygen atom and the two hydrogen atoms share electrons – each atom then has a full outer electron shell.

The covalent bond is a pair of electrons shared between two atoms. Covalent bonds are very strong. It takes a lot of energy to break them. The molecules themselves, however, attract each other only very weakly as each molecule is overall neutral.

> ✔ **Test yourself**
>
> 6 How is a covalent bond formed between two atoms?
> 7 Draw a dot and cross diagram showing the covalent bonds in the following molecules:
>   a) hydrogen chloride gas (HCl)
>   b) methane (CH$_4$)

Metallic bonding

Metallic bonds are formed between the atoms of a metal when ⊞ they come together to form the solid metal.

Solid metals have a structure consisting of a lattice (regular three-dimensional pattern) of positive ions through which a 'sea' of free electrons moves (Figure 18.4). The positive ions and the 'sea' of electrons interact to form metallic bonds.

The lattice/free electron model explains the high electrical and thermal conductivity of metals. The delocalised sea of electrons can move throughout the structure of the metal, forming an electric current. The positive ions are close together and bonded by metallic bonds. The structure can easily pass the vibration of hot particles from one particle to the next particle; and the free electrons can move faster as they are heated and transfer the energy from hot to cold throughout the structure – this explains why metals are good thermal conductors.

> **Key term**
>
> **Metallic bonds** Form when metallic positive ion cores get arranged in a lattice structure, through which a 'sea' of delocalised electrons can flow.

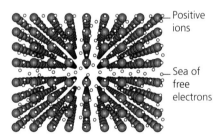

Figure 18.4 The metallic structure of copper

Positive ions

Sea of free electrons

Properties of metals

Metals have similar physical and chemical properties, which distinguish them from non-metals.

The general properties of metals are as follows:

▶ **Strong**: In science, a strong material is one that cannot be easily broken. The opposite of strong is brittle. Metals are generally strong, although some are much stronger than others. (Important exceptions: sodium and potassium.)
▶ **Malleable** and **ductile**: These are similar yet different properties. Malleable means that the material can be bent, hammered or squashed into different permanent shapes. Ductile means that the material can be stretched out into wires.
▶ **Hard**: Metals are difficult to dent and scratch. (Important exception: lead.)
▶ **High melting and boiling points**: An exception to this is mercury, which is a liquid at room temperature.

> ✔ **Test yourself**
>
> 8 Which two metal properties make copper a particularly suitable material for electrical wiring?
> 9 Why are metals good conductors of electricity?

- **Good conductors of electricity**: All metals conduct electricity to some extent, although their conductivity varies considerably.
- **Good conductors of energy by heating**: The metals have a high thermal conductivity. Those that are better conductors of electricity are also better conductors of energy by heating.
- **Shiny when polished**: This is sometimes described as being **lustrous**. The more reactive metals tend to build up a layer of oxide on their surface (e.g. rust on iron) which dulls them.
- **High density**: The way that metal atoms bond together tends to give them a high density.

▶ Properties of materials

The general properties of different materials ultimately depend on the bonding between the particles. However, there are some bulk properties of materials that are very important, and these can be used to explain the behaviour of many materials.

Key term

Bulk properties Properties of a large (hand-held) piece of the material.

Density, stress and Hooke's Law

Density

Density is a measure of how much mass (matter) is present in a given volume of a material – usually $1\,cm^3$ or $1\,m^3$. It can be calculated using the following equation:

$$\text{density} = \frac{\text{mass}}{\text{volume}}$$

Water has a density of $1\,g/cm^3$ or $1000\,kg/m^3$.

> ★ **Worked example**
>
> A block of packing material for a box has a volume of $0.4\,m^3$ and a mass of $4.8\,kg$. Calculate the density of the packaging material.
>
> Answer
>
> $$\text{density} = \frac{\text{mass}}{\text{volume}} = \frac{4.8\,\text{kg}}{0.4\,\text{m}^3} = 12\,\text{kg/m}^3$$

Stress

The **stress** exerted on an object is defined as the amount of force acting over the area of the object. Stress is an important measure of the physical properties of a material, such as strength. Materials that can withstand a lot of stress without changing shape are generally strong.

Stress can be determined by the equation:

$$\text{stress} = \frac{\text{force}}{\text{cross-sectional area}}$$

Force is measured in newtons, N, and cross-sectional area is measured in squared metres, m², so the units of stress are N/m².

> ### ★ Worked example
>
> Calculate the stress on a concrete pillar supporting a garden ornament with a weight (force) of 36 N. The pillar has a cross-sectional area of 0.004 m².
>
> #### Answer
>
> $$\text{stress} = \frac{\text{force}}{\text{cross-sectional area}} = \frac{36 \text{ N}}{0.04 \text{ m}^2} = 900 \text{ N/m}^2$$

Hooke's law

Figure 18.5 A rugby team scrummaging against their scrum machine

Figure 18.5 shows a rugby team scrummaging against a scrum machine. The scrum machine works by using a set of springs. As the players exert a force, that pushes against the pads and the springs inside the machine compress (get shorter). The larger the force exerted by the players, the larger the compression of the spring. The stiffness (or 'springy-ness') of the spring is measured using a property called the spring constant, k. The force applied, the change in length of the spring (extension or compression) and the spring constant are related by the Hooke's Law equation:

$$\text{force} = \text{spring constant} \times \text{extension}$$

> ### ★ Worked example
>
> The Bath Rugby scrum exerts a force against their scrum machine, causing a spring to compress by 0.125 m. If the spring constant is 40 000 N/m, calculate the force exerted on the spring.
>
> #### Answer
>
> $$F = kx = 40\,000 \text{ N/m} \times 0.125 \text{ m} = 5000 \text{ N}$$

> ### ✔ Test yourself
>
> 10 A gold ring has a mass of 5.79 g and a volume of 0.3 cm³. Calculate the density of gold in g/cm³.
> 11 A bungee rope supports a woman with a weight (force) of 600 N. The bungee rope has a cross-sectional area of 0.0004 m². Calculate the stress on the bungee rope.
> 12 During a gym session, a rugby player exerts a force on the spring inside a gym machine, extending it by 0.15 m. The spring constant of the spring is 8000 N/m. Calculate the force exerted by the rugby player on the spring.

▶ Testing material properties

The physical properties of material objects can be measured experimentally by using a few simple practical procedures. The properties, definitions and practical procedures are shown in Table 18.1.

Table 18.1 Properties of materials

Physical property	Definition	Practical procedure
Stiffness/flexibility	Stiffness is the resistance of an object to bending. (Flexibility is the opposite.)	An object is clamped to a desk at one end and weights are hung from the other end of the object. A ruler is used to measure how much the object bends at the weighted end.
Toughness/brittleness	Toughness is the ability of a material to absorb energy and stretch without fracturing. (Brittle materials do not stretch before they fracture.)	A notched standard sized material sample is clamped vertically at one end and a falling weight hits the other end. The weight is increased until the sample fractures.
Tensile (breaking) strength	Tensile strength is the maximum force that an object can take before it breaks.	One end of a thin standard sized material sample is clamped, and the other end is pulled with a newton meter until it breaks, and the breaking force is measured.
Hardness	Hardness is the resistance of an object to scratching.	The hardness of an object is measured by scratching it with the objects in a Moh hardness kit and seeing which objects will scratch it and which ones will not. A fingernail has a hardness of 2.5, a copper coin 3.5, a knife 5.5, a steel nail 6.5 and a masonry drill bit 8.5.
Density	The density of an object is its mass divided by its volume.	The mass of an object is measured with a balance and the volume either with a ruler, or by the displacement of water.
Durability	The durability of an object is its ability to withstand wear, pressure or damage.	The durability of a standard sized object can be measured by counting the number of times that the object can resist a force before breaking.
Shock absorption	Shock absorption is the ability of an object to absorb mechanical shock.	A simple shock test involves dropping an object from a measured height and observing if it breaks. The drop height is varied until the object breaks.
Thermal conductivity	Thermal conductivity is the ability of a material to allow the flow of energy by heating through it.	Identical metal rods made from different materials have a drawing pin fixed to one end with Vaseline grease. The other ends of the rods are heated in a flame. The rod with the drawing pin that falls off first has the highest thermal conductivity.
Electrical conductivity	Electrical conductivity is the ability of a material to allow the flow of electrical current through it.	The resistance of a standard sized piece of material is measured with a multimeter. The lower the resistance, the higher the electrical conductivity.

Key term

Durable Can withstand wear, pressure or damage.

⚙ Specified practical

Investigation of the thermal conductivity of metals

A student performed an experiment to investigate the thermal conductivity of different metal rods. She set up the experiment as shown in Figure 18.6.

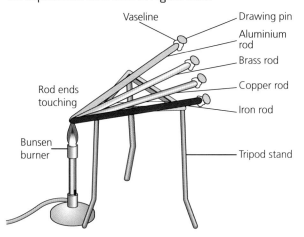

Figure 18.6 Apparatus to investigate the thermal conductivity of metal rods

The student used a stopwatch to time how long it took for each drawing pin to fall off its rod.

Results

Rod	Time for the drawing pin to fall off the rod (minutes and seconds)
iron	3 min 21 s
copper	1 min 12 s
brass	2 min 36 s
aluminium	1 min 48 s

Analysing the results

1 What is the order of thermal conductivity of the metal rods?
2 What is the dependent variable in this experiment?
3 Why is it important to have rods with the same length and thickness?
4 The hot metal rods can burn. What control measure should you take to ensure that you are not at risk from this hazard?

Types of materials

There are several main classes of materials, each with several properties common to all the materials in the class. The main classes of materials are:

- metals and alloys
- polymers
- ceramics
- composites

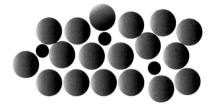

Figure 18.7 The atomic structure of an alloy

Metals and alloys

Mixtures of metals are called **alloys**. Sometimes designers and engineers need metal parts with very specific properties for very specific applications. In many cases, the natural metals do not have these properties, or the cost of natural metals for these applications would be uneconomic. In these cases, alloys are used. When the alloy is made from mixing two or more metals together, the atoms of each metal are different sizes; this tends to distort the regular structure of the metal (Figure 18.7). Generally, alloys are harder, stronger and less malleable than pure metals because it is more difficult for the layers of atoms to slide over each other.

Stainless steel is an alloy, which is used to make saucepans, cutlery and sinks. Other common alloys include aluminium alloys (used in bicycle and aircraft frames) and alloys of titanium (used in high performance aircraft frames and replacement hip joints). You do not need to know the compositions and properties of any alloys, but you may be asked to interpret information about them.

★ | Worked example

Steel is an alloy, primarily of iron and carbon, but its properties can be varied for different applications by varying the percentage composition of carbon or adding chromium. Table 18.2 shows some of the different types of steel.

Table 18.2 Different types of steel

Name of steel alloy	Composition	Properties	Uses
Cast iron	Iron 2% to 5% carbon	Hard Brittle Easily corrodes	Pots and pans Car parts Pipes
Mild (low carbon) steel	Iron 0.1% to 0.3% carbon	Tough Ductile Malleable Strong Easily corrodes	Car bodies Building lintels Food cans
High carbon steel	Iron 0.8% to 1.4% carbon	Very strong Very hard Durable Easily corrodes	Cutting tools Masonry nails Railway lines
Stainless steel	Iron 1.2% carbon 10.5% chromium	Hard Strong Durable Resistant to corrosion	Cutlery Surgical instruments Architectural features Exhaust systems

→

Polymers

Polymers are long chain covalent molecules where the atoms form repeating units. The bonds between the atoms are very strong, but the bonds between individual fibres are quite weak. Polymers are made mostly from atoms of carbon and hydrogen. Polymers are extremely important materials and synthetic polymers are called plastics. Polymers can be designed to have a huge variety of different properties and can be manufactured into any shape. Generally, polymers are less strong, less stiff and have lower melting points than metals, although there are very important exceptions (such as Kevlar©, which is very hard and strong, and is often used to make body armour).

The properties of polymers can be changed by adding cross-link molecules between the polymer chains, as shown in Figure 18.8. The cross-links prevent the polymers from easily sliding over each other. This makes cross-linked polymers less flexible, less stretchy, harder, tougher and with a higher melting point.

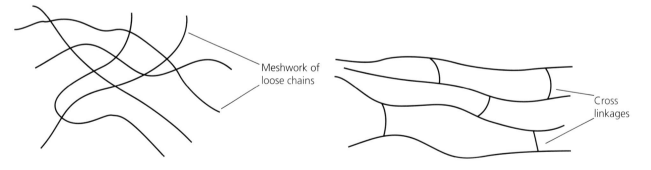

Meshwork of loose chains

Cross linkages

Figure 18.8 Structural differences between normal chain polymers and cross-linked polymers

Ceramics

Ceramics are a group of materials made from inorganic, non-metallic, raw material substances, such as clay. The wet clay is shaped into an object and then fired (heated to high temperature), which removes the water molecules from the clay. The particles in the clay form giant structures with strong covalent bonds between the particles.

Ceramic objects generally have these properties:

▶ hard
▶ **brittle**
▶ high melting point
▶ low thermal conductivity (good thermal insulator)
▶ high resistance to chemical attack.

Ceramics are useful materials for making common objects such as:

▶ tiles
▶ crockery (plates, bowls, cups, mugs)
▶ ornaments
▶ sinks and toilets

Composites

Composites are materials made from combinations of two or more materials from different classes. Material combinations are chosen so that the composite has the best properties of all the materials used to make it. Inside the composite, the different materials remain as separate materials, but are mixed together. A good example of a composite material is carbon fibre, (a composite of thin fibres of graphite and a resin glue) which is commonly used as panelling for aircraft and cars and is often used to make sporting equipment. Carbon fibre is an ideal material for making into lightweight but strong objects. Other common composites are:

▶ steel reinforced concrete
▶ chipboard (a mixture of wood chips and polymer resin).

Allotropes of carbon

Some covalent substances exist as giant structures, in effect, one large molecule. Graphite and diamond are examples of giant covalent structures (Figure 18.9). Diamond and graphite are different physical forms, or **allotropes**, of carbon. Both diamond and graphite contain covalent bonds between carbon atoms (Figures 18.10 and 18.11).

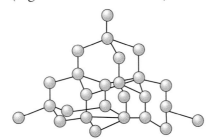

Figure 18.10 The structure of diamond

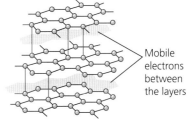

Mobile electrons between the layers

Figure 18.11 The structure of graphite

Properties of diamond and graphite

Graphite has very different properties to diamond, despite the fact that both are made solely of carbon atoms joined by covalent bonds (Table 18.3). This is because the atoms are arranged in different ways.

Figure 18.9 Diamond and graphite

Table 18.3 Physical properties of diamond and graphite

Physical properties of diamond	Physical properties of graphite
Transparent and crystalline – used as a gemstone in jewellery	Grey/black shiny solid
Extremely hard – used for glass cutting, small industrial diamonds are used in drill bits for oil exploration etc.	Very soft – used as a lubricant and used in pencils
Electrical insulator	Is a non-metal that conducts electricity. Graphite is used for electrodes in some manufacturing processes
Very high melting point, over 3500 °C	Very high melting point, over 3600 °C

Explaining the physical properties of diamond and graphite

The carbon atoms in the layers of graphite are held together by three strong covalent bonds, involving three of the four outer electrons. The fourth electron from each atom joins a delocalised system of electrons between the layers of carbon atoms, allowing it to conduct electricity along the layers. The hexagonal layers in graphite can slide over one another (because the bonds between the layers are very weak), giving graphite its slippery feel and lubricating properties. The carbon atoms are held together by strong covalent bonds, so that temperature has little physical effect on graphite and its melting point is high.

In diamond, all four of the outer electrons are involved in covalent bonding with four other carbon atoms, forming a tetrahedral structure. The result is a giant rigid covalent structure. This gives diamond its incredible hardness and high melting point, as a lot of energy is required to break down the lattice. There are no free electrons to conduct electricity.

Nanotubes, fullerenes, graphene, and carbon fibres

The fullerenes are a group of allotropes of carbon. They are made of balls, 'cages' or tubes of carbon atoms.

Nanotubes and carbon fibres

Carbon nanotubes are one type of fullerene (Figure 18.12). They are molecular-scale tubes of graphite-like carbon with remarkable properties. Carbon nanotubes are among the stiffest and strongest fibres, and they have amazing electrical properties: depending on their exact structure they can have higher electrical conductivity than copper – all in a tube about 10 000 times thinner than a human hair.

The covalently bonded hexagonal carbon sheets make carbon nanotubes incredibly strong, and the free electrons give them a high electrical conductivity. Most of the current applications of carbon nanotubes are based on their strength, such as car panels, bullet-proof vests, boat hulls and in epoxy-resins for bonding high performance components in wind-turbines and sports equipment.

Other fullerenes

Other types of fullerenes have different shapes to nanotubes. One example, Buckminsterfullerene ('buckyballs'), is shown in Figure 18.13.

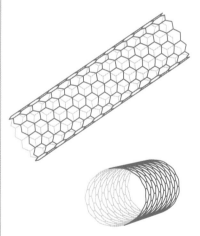

Figure 18.12 Carbon nanotubes are formed when graphite layers roll up into tubes

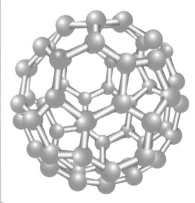

Figure 18.13 Structure of buckminsterfullerene

Buckminsterfullerene consists of a giant molecule of 60 carbon atoms in the shape of a ball. Other buckyballs have been made with 70, 76 and 84 carbon atoms. Their properties are generally similar to nanotubes and once again they are very strong, but because they have a closed structure, they can entrap other molecules so they can have additional applications such as carrying drugs for delivery to specific sites in the body.

Graphene

Graphene is similar to a single layer of the graphite molecule, in which the carbon atoms are bonded together in a hexagonal honeycomb lattice. It has many extraordinary properties.

- It is the thinnest material known (only one atom thick).
- It is the lightest material known (1 square metre of graphene weighs around 0.77 milligrams).
- It is 100 to 300 times stronger than steel, making it the strongest material discovered.
- It is the best conductor of energy by heating at room temperature and the best conductor of electricity known.

Graphene has a number of potential uses:

- It is very unreactive and so could be incorporated into paint, which would for example, prevent painted metal rusting.
- Its strength means that it is a potential replacement for Kevlar, the material currently used to make bulletproof vests.
- Its high electrical conductivity makes it an ideal material for making low energy light bulbs, lightweight but flexible display screens, and solar cells.

✔ Test yourself

18 Explain why graphite, despite being a non-metal, is a good conductor of electricity.
19 Explain why diamond is so hard.
20 What is the name of the group of materials that include nanotubes and buckyballs?
21 What would be the advantage of adding graphene to the paint used on the steel panels in car bodywork?

Choosing materials for a purpose

When a designer, engineer or scientist is choosing a material for a purpose, it is important that its physical properties are considered (Tables 18.1 and 18.2). These and other factors will also dictate if the material is suitable for the purpose:

Resistance to corrosion Some materials are used outdoors, where they are subject to the weather. A bench made from mild steel would be a good choice for indoor seating, but would quickly rust outside, particularly if the bench is by the seaside. Some materials are used in reaction vessels for chemicals or for preparing food, so need to be resistant to corrosion.

Biological inertness Materials used as surgical implants need to be biologically inert. This means that they do not interact with the surrounding body tissues or fluids.

Cost Gold is frequently used for important sporting trophies; however, gold is a very expensive metal, and so it would not be cost-effective to buy a solid gold trophy for a local children's competition.

Environmental impact Manufactured objects have to have an environmental impact assessment, which considers any potential impacts to the environment throughout the whole product lifecycle. This includes: obtaining the raw materials; energy expended during manufacture, use and recycling; environmental impacts during manufacture, use and recycling; and eventual disposal.

Sustainability Some materials make excellent fuels, but burning coal, oil and gas produces carbon dioxide gas which is a greenhouse gas and a main contributor to global warming. This makes fossil fuels unsustainable for the future. Some raw materials, such as wood, are also more sustainable than others. Wooden houses are more sustainable than houses built from steel and concrete.

In many cases, materials possess a combination of properties that makes them suitable for a purpose and the development of new materials has allowed their uses to change over time. Sports equipment, clothing, car and aircraft parts, and surgery have all benefitted from advances in the development of new materials. Tennis rackets, once made from wood, for example, are now made from graphene.

Chapter summary

- Metals, ionic compounds, and covalent substances are groups of materials, with similar properties.
- The physical properties of metals can be explained using the 'sea' of electrons/lattice of positive ions structural model.
- Ionic bonding involves the loss or gain of electrons. The resultant ions are held together by electrostatic forces.
- The charge on an ion depends on the position of the element in the Periodic Table.
- The structure of giant ionic substances explains the physical properties of ionic compounds.
- Covalent bonds are formed when atoms share electrons.
- Metallic bonding involves a lattice of positive ion cores and a 'sea' of delocalised electrons.
- The main classes of materials are metals and alloys, polymers, ceramics, and composites.
- Alloys are mixtures of two or more metals.
- Polymers are long chain organic compounds made from many repeating molecular units.
- Allotropes are different physical forms of the same element. The properties of the allotropes of carbon:

diamond, graphite, fullerenes, carbon nanotubes and graphene, can be explained in terms of their structure and bonding.

- Composite materials are a mixture of two or more different materials.
- The physical properties of a material that can be tested experimentally:

$$\text{density} = \frac{\text{mass}}{\text{volume}}$$

$$\text{stress} = \frac{\text{force}}{\text{cross-sectional area}}$$

- Hooke's Law: force = constant × extension
- Physical properties, including resistance to corrosion and biological inertness, as well as other factors such as cost, environmental impact and sustainability need to be assessed when choosing a material for a purpose.
- A combination of properties often makes a material suitable for a purpose, and the materials used for many objects have changed over time as new materials have been developed.

Materials for a purpose

▶ Practice exam questions

1 The use of a material is determined by its properties.

a) Draw lines linking the correct material to a description of its properties. [4]

Alloy   Brittle, hard, low electrical conductivity

Polymer   A mixture of two or more metals

Ceramic   A combination of materials from different material classes

Composite   Long chain organic molecule. Generally low melting point

b) Plastics are synthetic polymers. A plastic is needed to make the foam inside a baby's car seat. Some properties of some common plastics are given in the table below.

Plastic	Density (kg/m³)	Stiffness (GPa)	Tensile strength (MPa)
Polythene	960	1.1	25
PVC	1450	3.3	48
Polyurethane	1660	0.15	62
Polystyrene	45	0.007	0.4

Use the information in the table to answer the following questions.

i) State which plastic could be used to make the foam as flexible as possible. [1]

Give one reason for your answer. [1]

ii) State the name of the plastic that would make the foam as light as possible. [1]

iii) State the name of the plastic that would make the foam able to withstand large forces. [1]

Give one reason for your answer. [1]

2 Ionic bonding generally occurs between metals and non-metals. The electron structure of the metal lithium, and the non-metal fluorine is shown in the diagram.

 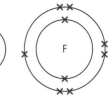

a) When lithium burns in fluorine gas, the ionic compound lithium fluoride is formed. Draw the dot-and-cross diagram for lithium fluoride. [4]

b) Fluorine gas is a covalently bonded molecule containing two fluorine atoms. Draw the dot-and-cross diagram for fluorine gas. [2]

c) Lithium is a shiny soft metal. Fluorine is a very pale-yellow gas. Lithium fluoride is a white crystalline ionic compound. Explain the following electrical tests performed on these materials. [3]

i) Fluorine gas does not conduct electricity.

ii) A solution of lithium fluoride dissolved in water conducts electricity.

iii) Solid lithium metal conducts electricity.

H

Producing food

The growing world population needs to be fed, and the production and processing of food requires a number of the scientific skills and techniques. We need to maximise food production of high-quality food. This requires modern farming practices, some of which have an impact on the environment.

▶ The importance of plants

Although it only happens in plants, all life on Earth depends on photosynthesis. It is the process that converts sunlight into food, both for plants and for the animals that form the food chains leading from those plants. It also produces oxygen as a waste product, allowing our atmosphere to support aerobic life. Scientists try to understand as much as possible about the process of photosynthesis, in the hope of being able to boost food production for the world's growing population.

What do plants need to survive?

To carry out photosynthesis and their other living processes, plants need certain materials from their environment (Figure 19.1).

The process of photosynthesis

Photosynthesis is a complex series of chemical reactions in the chloroplasts of plant cells. It can be summarised by:

$$\text{carbon dioxide} + \text{water} \rightarrow \text{glucose} + \text{oxygen}$$

For the process to work, four things are needed:

▶ **Carbon dioxide** Glucose is made of carbon, hydrogen and oxygen. The carbon dioxide provides the carbon and oxygen.
▶ **Water** This provides the hydrogen needed to make glucose. The oxygen from the water molecules is not needed and is given off as a waste product.
▶ **Light** This provides the energy for the chemical reactions in photosynthesis.
▶ **Chlorophyll** The green pigment in chloroplasts is chlorophyll, which absorbs the light to provide the energy for photosynthesis.

All of the chemical reactions involved in photosynthesis are controlled by enzymes, which are found in the chloroplasts of the photosynthesising cells.

Factors affecting the rate of photosynthesis

Photosynthesis makes food. The more photosynthesis there is occurring in a plant, the more food it makes. Commercial plant growers obviously want photosynthesis to happen as fast as possible in their plants, because that will mean their plants will grow quicker, or be bigger or healthier. If plants are grown in

FROM THE SUN
Energy transferred by light for photosynthesis

FROM THE AIR
Carbon dioxide – needed for photosynthesis
Oxygen – in the day, plants make more oxygen in photosynthesis than they need for respiration. At night, photosynthesis stops and plants need to obtain oxygen from the air.

FROM THE SOIL
Water – needed for photosynthesis and other living processes
Minerals – needed for a range of living processes; nitrates are needed to make proteins from the glucose made in photosynthesis

Figure 19.1 The needs of plants

greenhouses, the environmental conditions can be controlled to maximise photosynthesis. We know that the external factors needed for photosynthesis are light, carbon dioxide, water and a suitable temperature:

- **Light** Increasing light intensity boosts the rate of photosynthesis, but only up to a point. The amount of chlorophyll in a plant is fixed at one time. If the light intensity is greater than the chlorophyll can absorb, any further increase in intensity has no effect.
- **Carbon dioxide** Increasing carbon dioxide levels will increase the rate of photosynthesis up to a certain level, and then increasing it further has no effect. Once the chloroplasts have all the carbon dioxide they can use, there is no benefit in increasing it.
- **Temperature** The chemical reactions in photosynthesis are all controlled by enzymes, and the effect of temperature on the rate of photosynthesis is due to the effect of temperature on those enzymes. Raising the temperature up to about 40 °C is beneficial, as long as the plant doesn't dehydrate in the process. As the temperature gets higher still, it destroys (denatures) the enzymes and photosynthesis stops.

If the plant has enough water to survive, it will have enough for photosynthesis. Extra water will not increase the rate.

Limiting factors

In any set of circumstances, one factor, known as the limiting factor, sets the rate of photosynthesis. In different conditions, any of the factors listed above – light, carbon dioxide or temperature – can be the limiting factor. You can tell if a factor is limiting by increasing it. If the rate of photosynthesis also increases, then the factor was limiting.

Processing glucose after photosynthesis

Just like animals, plants need a variety of nutrients. The difference is that they must make the nutrients themselves (apart from minerals, which are absorbed from the soil). They make a variety of carbohydrates and proteins. They have less need for lipids, although some seeds do use oils as a food store. Carbohydrates and lipids can be made from glucose because they contain the same chemical elements (carbon, hydrogen and oxygen). Proteins include nitrogen which is absorbed from the soil in the form of nitrates.

The main ways in which glucose is used in plants after it is formed in leaves are shown in Figure 19.2.

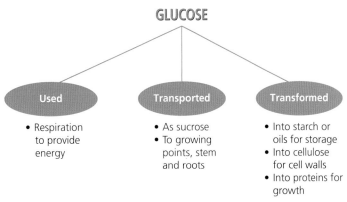

Figure 19.2 The fate of glucose made in photosynthesis

The structure of a leaf

A leaf is a complex organ, with features that make it well adapted to its function of carrying out photosynthesis. Light is absorbed by green chlorophyll, stored in chloroplasts in the leaf cells, and the structure of the leaf ensures that the cells containing the chloroplasts receive the water and carbon dioxide they need for photosynthesis. The internal structure of a leaf is shown in Figure 19.3.

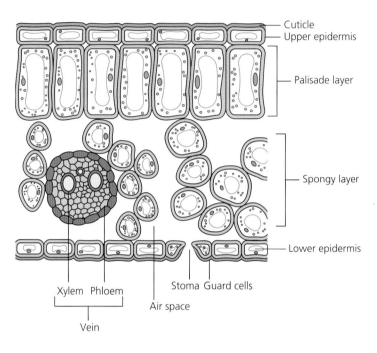

Figure 19.3 The internal structure of the leaf

The functions of each of the structures in relation to photosynthesis are given in Table 19.1.

Table 19.1 Structure and function of leaf structures

Structure	Function
Cuticle	Waxy, waterproof layer that reduces water loss – it is transparent, allowing light through to the lower layers of cells, which contain chloroplasts
Palisade layer	The cells are packed with chloroplasts for photosynthesis
Spongy layer	Contains large air spaces, allowing carbon dioxide to reach this layer for photosynthesis, but the cells here also contain chloroplasts for photosynthesis
Vein	Contains xylem (brings water to leaf) and phloem (transports sugar away)
Guard cells	Open and close the stomata, allowing carbon dioxide to enter or preventing water loss

To let carbon dioxide in for photosynthesis, the leaf has pores called stomata (singular: stoma) that are open to the atmosphere. It is impossible to let carbon dioxide in without also allowing water to escape, and water is a valuable resource. In the daytime, water loss is bound to happen, but at night, when no photosynthesis can occur, the loss of water would be a waste. To reduce this water loss, the guard cells around each stoma can change shape and cause the stomata to close.

Food for the future

Test yourself

1 Name the four factors needed for photosynthesis.
2 If you increase the light intensity shining on a plant, what will happen to the rate of photosynthesis?
3 A horticulturalist increased the level of carbon dioxide in her greenhouse by installing a burner outside and piping the carbon dioxide produced into the greenhouse. The yield of plants in the greenhouse increased. What conclusions would you draw from this?
4 Photosynthesis alone cannot supply the plant with protein. What else is necessary?
5 Suggest a reason why the palisade layer is found in the top half of the leaf.

Specified practical

Investigation of factors affecting photosynthesis

Light is one of the factors which affects the rate of photosynthesis. In this investigation a green plant named Canadian pondweed (*Elodea*) produced bubbles of oxygen as a result of photosynthesis.

The effect of light intensity on the number of bubbles produced was investigated. The apparatus used is shown in Figure 19.4.

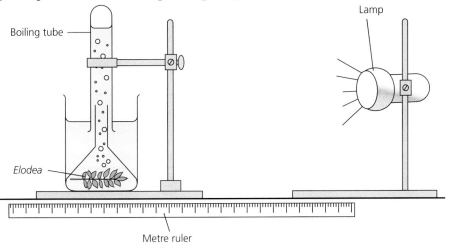

Figure 19.4

Procedure

1 The *Elodea* was placed in a beaker containing 200 cm³ of water.
2 One spatula of sodium hydrogen carbonate was added to the water.
3 Three small pieces of plasticine were stuck to the rim of the funnel and it was placed upside down over the plant.
4 A test tube was filled with water and carefully placed over the end of the funnel with the end under the water and was clamped in place.
5 The lamp was placed 5 cm away from the apparatus.
6 The number of bubbles of oxygen produced in one minute was recorded.
7 The experiment was repeated with the lamp 10 cm, 15 cm, 20 cm, 25 cm and 30 cm from the apparatus.

Analysing the results

1 Explain the reason for the following:
   a) The addition of sodium hydrogen carbonate to the water.
   b) The use of the plasticine.
2 Suggest one source of inaccuracy in this experimental design.
3 In an improved design, a sheet of glass was placed between the lamp and the beaker. The glass allows light, but not heat, to pass through. Suggest why this design is an improvement over the first one.
4 A student suggested that if the experiment was repeated 5 times for each distance, the repeatability of the result would be improved. State why this is incorrect.

Plant mineral requirements

Plants need a variety of minerals for healthy growth, but three of the most important are nitrogen, potassium and phosphorus. If a plant lacks one of these, it shows poor growth and specific symptoms.

▸ Lack of nitrogen (in the form of nitrates) causes generally poor growth because this mineral is needed to make proteins for new cells.
▸ Lack of potassium is characterised by yellowing of the leaves.
▸ Lack of phosphorus (in the form of phosphates) results in poor root growth.

General-purpose fertilisers are often referred to as NPK fertilisers because they contain nitrogen (N), phosphorus (P) and potassium (K).

Different farming methods

Farming methods in the UK fall into two different groups.

▸ **Intensive farming** which aims to maximise food yields. The method uses high yield crops, together with fertilisers and pesticides. Highly efficient machinery is used.
▸ **Organic farming** focuses on 'natural' methods, avoiding the use of chemical fertilisers, pesticides, growth hormones and livestock feed additives. It uses crop rotation and animal manure to maintain soil fertility, and biological methods of pest control.

The differences between intensive and organic farming are summarised in Table 19.2.

Key term

Crop rotation The practice of growing a series of different crops in the same area in successive growing seasons, avoiding exhaustion of soil nutrients; peas and beans actually add nitrates to the soil.

Table 19.2 Differences between intensive and organic farming

	Intensive farming	Organic farming
Main aim	Producing as much food as possible, as cheaply as possible	Producing high quality food in ways that benefit humans and the environment
Costs	Costs are kept to a minimum, so the food produced is as cheap as possible	Costs are considered, but the method is less efficient than intensive farming, so the food produced costs more
Use of chemicals	Chemical fertilisers and pesticides are used routinely	Chemical fertilisers are replaced with natural ones (animal manure or plant waste); pesticides are not used
Animal welfare	Animals may be kept indoors where their environment can be carefully controlled. The conditions are chosen so the animal grows quickly and may involve very cramped conditions	Animals are kept outside (except in bad weather or at night) and are not crowded together
Food additives	Growth hormones and antibiotics may be added to animal feed. The antibiotics are a precaution against disease	Not used
Pest control	By chemical pesticides	Biological control

Key term

Biological control The use of natural predators to control pests.

Advantages of different farming methods

The advantages of intensive farming are:

▸ The yields are high.
▸ The food is produced more quickly than on organic farms.
▸ The food is cheaper to produce and therefore prices are lower.

The advantages of organic farming are:

▸ No possible pollution of the environment by pesticides and chemical fertilisers. Pesticides can kill non-pest organisms, and fertilisers can cause eutrophication in streams and lakes (see page 104). But note that organic farms can cause eutrophication due to leaching of manure.
▸ Does not use antibiotics (which are a potential cause of 'superbugs' – see page 131).
▸ Does not use animal feed additives which can end up in the food consumed by humans.
▸ Animals generally live in better conditions on organic farms.

Ethics in farming

The main ethical issue in farming results from the conflict between profit (and cheaper food) and animal welfare. Animal welfare can be improved by keeping stock in more humane conditions, but some people believe we should never kill animals for food.

> ✓ **Test yourself**
>
> 6 Explain how crop rotation benefits the growth of plants.
> 7 Why do some farmers add antibiotics to animal feed?
> 8 Why is it untrue to say that organic farmers do not use fertiliser?
> 9 State one form of pollution that might result from organic farming.

Hydroponics

Plants are usually grown in soil, which varies in mineral content and types of bacteria, for example. Now, there is a growing trend in intensive farming of growing plants in facilities where the soil has been replaced with a mineral solution that is pumped around the plant roots (Figure 19.5). This system is called **hydroponics** and has a number of advantages compared to growing in soil.

▸ There is complete control over nutrient balance – the exact amount of nutrients can be added.
▸ Plants can be grown anywhere, even in areas where the soil is infertile.
▸ Some pests and pathogens come from soil. These are not a problem in hydroponics, so there is less pesticide use.
▸ Hydroponics uses less space for growing.
▸ Harvesting is easier as soil does not need to be cleaned off, and fine roots are not damaged.
▸ No weeding is necessary (soil normally contains seeds and spores which can grow into weeds).

Figure 19.5

Genetically modified crops

Scientists can extract functioning genes from one organism and put them into the chromosomes of another. They can also 'swap' one gene for another. The introduction of genes into food plants is becoming more common and is known as genetic modification (GM). In the 1980s, a commercial genetically modified (GM) potato crop was developed, which was modified so it made its own built-in insecticide. The insecticide was an insect poison normally produced by a type of bacterium that lives in the soil. The gene for the poison's production was transferred to potato plants, which made the plant resistant to insect pests.

Weeds compete with crops. For many years farmers have attempted to get rid of weeds using chemicals called herbicides. However, selective herbicides that kill only weeds and not the crop plants are difficult to produce. A herbicide-resistance gene can be taken from a bacterium that normally grows in soil and transferred into a plant such as soya. Herbicides can then be used to kill weeds in soya bean fields and so improve yields.

Some people do not like the idea of GM crops, and there are both advantages and disadvantages connected with their development.

The case for GM:

- ▶ Crops can be tailor-made to suit varied conditions around the world, potentially providing greater nutritional value and a higher income for farmers.
- ▶ Energy-producing biofuel crops could save natural resources and so conserve the environment.
- ▶ Increased yields could solve some food shortages around the world.

The case against GM:

- ▶ GM crops could reduce developed countries' reliance on crops from developing countries, resulting in loss of trade and severe economic damage for the developing countries.
- ▶ It is difficult to stop the pollen from outdoor GM crops from pollinating other nearby crops. People who do not want to grow GM crops (such as organic farmers) may end up with modified genes in their crop.
- ▶ The companies that develop GM crops have patents granting legal ownership, meaning only they can distribute seeds. They control the seed price, which may be too expensive for farmers in poorer countries.
- ▶ If crops are made resistant to herbicides, they could themselves become a pest species outside their environment (e.g. in gardens).

GM crops in the UK

There are currently no GM crops grown for sale in the UK. Field trials for research are allowed, and in 2021 the Government reduced the costs and regulations involved in setting up these trials. Research examples of genetic modifications which have already been developed and used in other countries, and write a report which answers the following questions:

1 What are the advantages and disadvantages of GM crops?
2 What benefits might consumers in the UK gain if GM crops were grown here?
3 State, with justifications, whether you consider that GM trials in the UK ought to be extended.

▶ Selective breeding in plants

Selective breeding is a process by which plants with beneficial features (caused by a particular allele) are deliberately bred together to produce large numbers of plants with this feature.

For example, it is an advantage to farmers if cropping apples stay on the tree so they can be picked, rather than falling off and becoming damaged. If a farmer wants to breed apple trees which retain their fruit:

1 They select the trees which retain the most fruit, and artificially cross-breed them.
2 The seeds produced in step 1 are grown into new trees.
3 From those trees, the farmer again selects the ones that have the best fruit retention and uses them to breed the next generation.
4 This process is repeated for many generations, and fruit retention gradually improves over this time.

There are clear advantages to selective breeding, but there are also disadvantages.

Advantages of selective breeding

Plants can be bred for:

▶ resistance to pests and diseases
▶ better yields
▶ greater nutritional value
▶ loss of damaging characteristics.

Disadvantages of selective breeding

Inbreeding causes:

▶ genetic variation to decrease, meaning the whole plant population may be susceptible to a given disease.
▶ loss of genes, making it more difficult to produce new varieties in the future.
▶ increase in the risk of genetic disease.

Key term

Inbreeding The breeding of individuals which are closely related and so share many similar alleles.

10 Explain why less pesticides need to be used in hydroponics facilities.
11 Explain how inserting herbicide resistance genes into a crop can increase the yield.
12 Organic farmers do not use GM crops. Suggest why an organic farmer might not want any GM crops on neighbouring farms.
13 Herbicide resistance in crops is a good thing. Suggest one reason why it might be better to use genetic modification to achieve this, rather than selective breeding.

Chapter summary

- Plants require certain materials to support life processes.
- The structure of a leaf suits its purpose to carry out photosynthesis.
- Photosynthesis is the process whereby green plants use chlorophyll to absorb light energy and convert carbon dioxide and water into glucose, producing oxygen as a by-product.
- Light, water, carbon dioxide and a suitable temperature H are required for photosynthesis and its rate is affected by light intensity, carbon dioxide and temperature.
- A limiting factor restricts the rate of photosynthesis.
- Glucose produced in photosynthesis may be respired to provide energy, converted to starch or oils for storage or used to make cellulose and proteins which make up the body of plants.
- Plants require certain nutrients for healthy growth. Lack of nitrates results in poor growth, deficiency of potassium results in yellowing of the leaf and deficiency of phosphate results in poor root growth.
- NPK fertilisers are used to promote growth in plants.

- There are two methods of farming: intensive and organic farming.
- Pesticides and fertilisers have an impact on the environment.
- There are differences of opinion on the ethics of these methods of food production.
- Food products can be grown in controlled environments to increase productivity.
- Hydroponics is a method of growing whereby soil is replaced by a mineral solution pumped around the plant roots.
- Artificial transfer of genes from one plant species to another (genetic modification) can be used to increase crop yield or improve product quality.
- There are potential disadvantages and issues involved in using GM crops.
- Selective breeding in plants can be used to produce desirable traits, but there are some disadvantages, such as a reduction in variation and increased susceptibility to disease.

Food processing and spoilage

The food industry requires the use of scientific skills and techniques to ensure that food is processed in ways that are safe and appealing to consumers. Increasing the shelf-life of foods has safety implications and can reduce waste. Taste is also important because if safety measures compromise taste, consumers will not buy the product. Microorganisms play a role in both food production and spoilage.

▶ Food production

Microorganisms are used in the production of bread, wine, beer, yoghurt and cheese.

Bread

Yeast, a microscopic fungus, has been used in breadmaking since the time of ancient Egypt, and maybe longer. Yeast in the bread mix breaks down the starch in flour into sugars, which are then respired anaerobically to form carbon dioxide and ethanol. Anaerobic respiration was covered in Chapter 1 Cells and respiration, but in yeast it forms carbon dioxide and ethanol rather than lactic acid. The carbon dioxide causes the bread dough to 'rise' (inflate) before baking, giving the bread a light texture. The sugars produced from the starch and the ethanol affect the taste of the bread.

Alcoholic drinks

As yeast produces alcohol (ethanol) when it respires anaerobically, it is also used in the production of alcoholic drinks. The variation in the drinks is a result of the food material given to the yeast (e.g. grapes in the case of wine, barley for beer).

Beer production

Beer production involves the following stages.

- ▶ **Malting** The barley seeds are soaked and allowed to germinate, then dried to stop growth. It is important that the seeds have a plentiful supply of air for germination to occur. The germination activates enzymes in the barley, which will break down starch and proteins in the mashing process.
- ▶ **Mashing** The malted barley is ground up and mixed with warm water. The temperature is adjusted at various points to favour the action of certain enzymes in the grain, which eventually convert starch to maltose and glucose (which the yeast will use as food).
- ▶ **Flavourings** The most common flavouring for beers is hops.
- ▶ **Boiling and cooling** The liquid (called 'wort') is boiled and then cooled. The boiling will destroy most microbes.

- **Fermentation** Yeast is added. There is oxygen in the wort and initially the yeast respire aerobically, but when the oxygen runs out anaerobic respiration (fermentation) and the formation of alcohol begins.
- **Barrelling or bottling** Fermentation ends when the maltose runs out and/or when the build-up of alcohol kills the yeast. The beer is then placed into barrels or bottles.

Yoghurt

Yoghurt is formed by the action of bacteria on milk. Lactose in the milk is converted to lactic acid by the bacteria. The change in pH curdles the milk and alters its taste. The processing consists of the following stages.

- **Equipment is sterilised.** It is important that only the yoghurt bacterial cultures grow, so all equipment used must be sterilised.
- **Pasteurisation.** The milk is heated to kill bacteria already present in it. There is more about pasteurisation below.
- **Bacterial cultures added.** The milk is cooled to 46 °C and a bacterial culture is added. This temperature promotes the growth of the bacteria and the formation of lactic acid. It takes about 4 hours for the yoghurt to form.
- **Flavourings added.** At this stage fruit and/or flavourings may be added. The yoghurt is then packaged.

Cheese

Cheese is also produced using cultures of bacteria. In some cheeses, fungi are also added. The processing is as follows.

- **Bacteria and rennet are added**. Bacterial cultures are added as in the making of yoghurt to produce lactic acid, but in cheese making rennet (containing the enzyme rennin) is also added to coagulate the milk, and form curds (solid) and whey (liquid). The milk used may or may not be pasteurised.
- **Whey is drained off**. The whey has no further use in the cheese making process.
- **The curds are compressed**. This forms the cheese.
- **Maturation.** The cheese is left for a time to mature, which improves its flavour and consistency. For blue cheeses, fungi are added which grow through the cheese to make blue 'veins'.

Optimum conditions for bacterial growth

It is important to understand the optimum conditions for bacterial growth. These must be provided if bacteria are used in food production and denied if you wish to slow bacterial growth to preserve food. These conditions are:

- **A suitable temperature**. Most bacteria thrive at warm temperatures (30–40 °C). Higher temperatures kill many (but not all) bacteria.
- **Moisture.** Like all living organisms, bacteria require water.
- **A food source**. To get energy, bacteria must have a source of food.

In food production methods that use bacteria, these conditions must be provided and monitored to ensure bacterial growth. The temperature and food source used can influence the final product.

Pasteurisation and the processing of milk

Pasteurisation is a process widely used in the food industry as a food safety measure. It is named after the famous scientist Louis Pasteur, who discovered it. Most of the milk we drink is pasteurised. This involves heating milk to 72 °C for at least 15 seconds. It is then rapidly cooled to less than 3 °C and packaged. At a temperature of 3 °C any bacteria which gets into the milk during the packaging process is unable to grow.

Pasteurisation does not kill all the bacteria in the milk (that is done in the production of sterilised and UHT milk), but it does kill all the known pathogens that might be present.

In addition to pasteurisation, milk is normally **homogenised**. Naturally, when milk is left to stand, it would separate, with cream rising to the top, above a thinner milk. To avoid this, milk is usually homogenised. This involves pumping the milk at high pressures through narrow tubes. The fat globules in the milk are broken up into smaller droplets so that they stay suspended in the milk (referred to as an emulsion).

Pasteurisation is also used in the production of beer and fruit juice.

> **Key term**
>
> Emulsion A mixture of two or more liquids, in which one is present as microscopic droplets and is distributed throughout the other.

Skimmed and semi-skimmed milk

Milk is sold in several varieties according to its fat content. **Full fat milk** comes from breeds of cattle that produce milk with a high fat content, with a distinct cream layer at the top. **Whole milk** is the 'normal' milk whose fat content has not been adjusted in any way. Before the milk is homogenised, the fat rises to the surface. Some of it can then be skimmed off to give **semi-skimmed milk,** or most of it can be removed to produce **skimmed milk**.

Test yourself

1 Why is it necessary to restrict oxygen levels in equipment when making wine or beer?
2 What role do hops play in the making of beer?
3 Why is yoghurt formation carried out at 46 °C rather than room temperature?
4 Explain the difference between pasteurisation and sterilisation.

▶ Food preservation

Food spoilage is caused by bacteria and/or fungi. Conditions that favour the growth of microorganisms will speed up food spoilage and if those conditions are absent, spoilage will be slowed or even stopped.

Slowing the growth of bacteria

With the exception of fresh food, most of the foods we buy have been treated in some way to slow or stop the growth of bacteria. This can be done in many ways, both by the manufacturers and by the customer once the food has been purchased.

Refrigeration and freezing

Cold temperatures reduce the growth of bacterial populations, allowing food to be kept fresh for longer. Refrigeration slows the growth of bacteria but does not stop it altogether. The useable life of food is extended but not indefinitely. Freezing makes the temperature so low that bacterial growth virtually stops, although the bacteria are not killed. Food keeps for much longer in a freezer than in a fridge. Once the food is thawed, bacterial growth will start again. Freezing is not suitable for all foods however, as the ice crystals that form may change the food in an unacceptable way.

Heating

We have already seen that heating is used in the pasteurisation of milk. Extreme heat kills bacteria and if the food is then packaged before new bacteria can enter, food products can last for a very long time, provided they are not opened.

Drying and salting

These preservation techniques operate on similar principles. Bacteria cannot survive without water. Drying removes water, but so does salting, as the salt absorbs water. Dried foods like raisins and dried herbs have a long shelf life. Salting of meat and fish used to be common before people had refrigerators, and salt is still added to many processed foods as a preservative. Salt affects the taste of the food, however.

Smoking

Hot smoking, as the name implies, exposes food to hot smoke. It preserves food in three ways: the heat kills bacteria; chemicals in the smoke act as preservatives because they also kill bacteria; the heat dries the food so that bacteria are denied access to water. Smoking affects the taste of the food and is only used with foods where the change of taste is considered desirable.

Pickling

Pickling involves putting the food in vinegar. Vinegar is an acid and bacteria cannot tolerate its low pH.

▶ Food hygiene

Safety precautions must be taken in any area that is used for food preparation, whether a small kitchen or a factory producing food products. Basic precautions are listed below.

- ▶ **Personal hygiene** Anyone in or around food preparation areas must wash their hands and wear protective clothing as appropriate (coats, gloves, hair coverings).

▶ **Detergents and disinfectants** Food preparation areas should be cleaned with detergent and then disinfectant. Detergent removes dirt on which bacteria could breed but does not effectively remove the bacteria themselves. Disinfectant should be used on food preparation surfaces. Disinfection kills bacteria but does not affect bacterial spores, so it is not as strong a measure as sterilisation.

▶ **Sterilisation** Some foods are sterilised before packaging. Any equipment used must itself be sterilised, to avoid contamination. This is usually done by heating (for example with steam).

▶ **Waste disposal** Waste food is a potential food source for bacteria and pests, so must be disposed of regularly.

▶ **Pest control** Food attracts pests, such as insects, mice and rats. Products for storage must be checked over when they arrive. Food establishments must have pest prevention methods already in place, with a fully pest-proof building. Measures include fly screens, wire mesh in air vents, metal grates over drains and trimmed back external vegetation. If rats or mice are found, the facility must close and no food preparation should be carried out until the pest infestation is dealt with. All food that may have been contaminated must be thrown away and all storage areas, food utensils and surfaces must be sterilised.

Cross-contamination

Cross-contamination occurs when bacteria is spread between food, surfaces and equipment. The following measures should be taken to avoid it.

▶ Raw food must not be allowed to come in contact with cooked food. The two types of food should be stored separately.

▶ Different equipment should be used for raw and cooked food.

▶ Appropriate protective clothing should be worn to prevent contamination.

 # Food poisoning

Food poisoning is caused by eating food contaminated with pathogens. Symptoms can range from mild to severe, and commonly include:

▶ nausea

▶ vomiting

▶ diarrhoea

▶ stomach cramps

▶ fever

▶ aching muscles

▶ chills.

Very severe cases of food poisoning can lead to hospitalisation or even death. The pathogens involved are usually bacteria, commonly *E. coli* or species of *Campylobacter* or *Salmonella*. These pathogens produce toxins and cause symptoms when present in large numbers. As the name implies, food poisoning is caused by eating contaminated food due to failures in food hygiene practice, either in the home or at a restaurant, shop, or food production facility.

Potential impact of food contamination

In a 2019 survey by the Food Standards authority (FSA), 47% of respondents reported that they had experienced food poisoning at some time in their lives, an increase on previous years. Researchers have estimated there are 180 deaths per year in the United Kingdom caused by foodborne diseases from 11 pathogens and the FSA estimates that about 2.4 million cases of foodborne illness occur every year in the UK.

Four species of bacteria (*Campylobacter*, *Clostridium perfringens*, *Listeria monocytogenes* and *Salmonella*) and one virus (norovirus) are responsible for most of the deaths.

The situation is worse in some parts of the world, with an estimated 600 million people falling ill because of eating contaminated food and a death toll of 420 000 per year.

Food contamination also has economic costs – people with food poisoning need medical care, productivity is affected by staff absence and people can be made unemployed if businesses close because of food contamination issues.

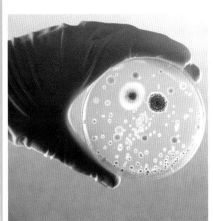

Figure 20.1 Agar plate showing bacterial colonies

▶ Growing microorganisms

Microorganisms are very small and difficult to see. To study them, scientists grow them in large numbers. This is often done on a type of jelly called agar, which has nutrients added to feed the microorganisms, in a plate called a Petri dish. Bacteria grow very quickly, and each bacterium that lands on the agar soon grows into a circular patch called a colony, which can be seen with the naked eye. Colonies of bacteria are shown in Figure 20.1.

Because each bacterium grows into a colony, by counting the colonies we know how many bacteria were originally put onto the plate in the original sample.

It is important that bacterial cultures are not contaminated by other microorganisms. This means that sterile conditions must be used:

▶ All surfaces are wiped with disinfectant before the procedure starts.
▶ The Petri dishes, nutrient agar and other culture media and equipment are sterilised before use.
▶ Bacteria are transferred to the Petri dish using a wire loop, which is sterilised by heating in a Bunsen flame.
▶ The lid of the Petri dish is sealed with adhesive tape to stop microorganisms from the air contaminating the culture.

In addition to counting colonies, bacteria growth can also be assessed in liquids by measuring turbidity (cloudiness). Bacterial growth creates cloudiness in liquids that would normally be clear.

Test yourself

9 Chicken meat contains *Salmonella* bacteria, beef does not. Suggest why beef can be eaten undercooked or even raw, whereas chicken must always be thoroughly cooked.

10 State **two** possible problems caused by food poisoning, **apart from** the unpleasantness caused to individuals by sickness.

11 Suggest why turbidity is not used to assess bacterial content of milk.

12 Sometimes, bacterial cultures need to be diluted before being grown on agar plates and counted. This is to stop the colonies overlapping. State why overlapping colonies are a problem.

Chapter summary

- Bacteria, yeast and other fungi are used in food production (bread, wine, beer, yoghurt and cheese).
- The processing of yoghurt, cheese and beer has several stages.
- There are optimum conditions for growth of bacteria (suitable temperature, moisture, food source) and this has significance in food production.
- Pasteurisation is a process which slows microbial growth in foods including beer, milk and fruit juice.
- Pasteurisation of milk is done by heating sufficiently to kill some pathogens.
- Semi-skimmed and skimmed milk are produced by removing fat from the raw milk.
- Milk is homogenised by pumping it at high pressure through narrow tubes, affecting the size of fat globules in milk and resulting in an emulsion.
- Food spoilage is due to bacterial and/or fungal action and may be accelerated by storage conditions.
- The growth of bacteria is slowed down or stopped by refrigeration, freezing, heating, drying, salting, smoking and pickling (lowering pH).
- Food preparation areas must be kept free of bacteria by means of personal hygiene, disinfectants, detergents, sterilisation, disposal of waste and control of pests, such as insects, mice and rats.
- Cross-contamination of food must be prevented during food preparation in food production facilities and in the home.
- Food poisoning is caused by the growth of microorganisms, usually bacteria, and by the toxins they produce when they grow (*Campylobacter sp.*, *E.coli, Salmonella sp.*).
- There are common symptoms for food poisoning (stomach pains, vomiting, diarrhoea).
- The growth of microorganisms can be assessed by colony counts and turbidity.
- Contamination of food products with bacteria has many potential impacts.

Practice exam questions

1 The leaves of a plant carry out photosynthesis. The structure of a leaf is shown in the diagram.

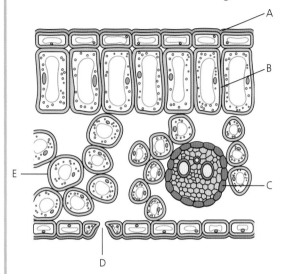

a) Identify structures A–E, using labels from the following list:

Stoma, upper epidermis, lower epidermis, palisade cell, air space, cuticle, vein, chloroplast. [5]

b) Write the word equation for photosynthesis. [3]

c) State three factors which affect the **rate** of photosynthesis. [3]

2 Scientists and farmers can improve crops by genetic modification and selective breeding.

a) Which of the following is a potential advantage of genetic modification? [1]

i) Crops will be able to cross-pollinate with other plants.

ii) Crops can be made resistant to herbivores.

iii) Crops can be made resistant to disease.

iv) Crops can produce seeds for a seed bank.

b) State **two** possible disadvantages of genetically modified plants. [2]

c) A farmer wants to develop a strain of strawberry which produces larger fruit. Outline how the farmer could use selective breeding to achieve this. [4]

3 a) Explain the purpose of each of these steps taken during the brewing of beer.

i) The barley seed is germinated. [1]

ii) Warm water is used to soak the crushed grain and break down starch and protein. [1]

iii) Hops are added to the liquid 'wort'. [1]

iv) The wort is boiled and cooled before yeast is added. [2]

v) The brewing process with yeast is done at low oxygen levels. [2]

b) Yeast is also used in bread making. Explain the role that yeast plays in the making of bread. [3]

4 The graph shows reported cases of food poisoning in Wales between 1992 and 2011. Note that each year's data spans two calendar years (e.g. 1992-3. 1993-4 etc.).

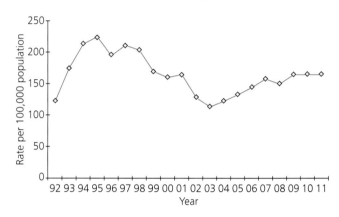

a) In which year was the highest rate of increase? [1]

b) In which year was the highest rate of decrease? [1]

c) There was a general rise in cases between 2003 and 2011. Which of the following statements offers a **possible** explanation for this rise? You can give more than one answer. [1]

i) People were going out to eat at cafes and restaurants more often.

ii) There were more checks on food preparation firms than in previous years.

iii) People were more likely to go to the doctor and report mild food poisoning symptoms than before.

iv) Food hygiene standards had declined.

The table shows how food poisoning cases in Wales were distributed throughout the year in 2011.

Quarter	Number of reported cases	% of yearly total
Jan–March	877	17
April–June	1353	26
July–Sept	1624	32
Oct–Dec	1257	25

Source: Public Health Wales

d) Suggest a reason why the January–March quarter shows the lowest incidence of food poisoning. [1]

e) Which of the following is not a symptom of food poisoning? [1]

i) vomiting   iii) stomach cramps

ii) diarrhoea   iv) blurred vision.

21 Scientific detection

Analytical chemistry is a branch of science in which samples are taken and tested to identify their components, and sometimes to find out the amounts of the substances in the sample. Samples should be representative, so they can be used to predict the properties of the whole material.

Analytical chemistry can be carried out in everyday settings by non-scientists, such as the staff who do drugs testing of luggage in airport security, or in high technology manufacturing plants by skilled scientists who, for example, assess the quality of manufactured food.

▶ Measuring substances

Each element on the Periodic Table has a symbol with its atomic number and **relative atomic mass**. When atoms join together to make molecules and compounds, the mass of these substances can be calculated by the sum of the relative atomic masses for each atom in the formula. The symbol for relative atomic mass is A_r. There is no unit for relative atomic mass, it is just a number.

> **Key term**
>
> **Relative atomic mass** The weighted average mass of an atom considering the isotopes available.

> ✔ | **Test yourself**
>
> 1 Name the element with the relative atomic mass of 1.
> 2 Give the symbol of the element with the relative atomic mass of 12.
> 3 What is the relative atomic mass of oxygen?
> 4 What is the relative atomic mass of N?

> **Key term**
>
> **Relative molecular mass** The sum of the relative atomic masses of all the atoms in a molecule.

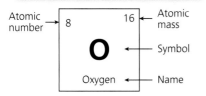

Figure 21.1 Close up of an element on the Periodic Table

Water is a compound of hydrogen and oxygen with the formula H_2O, so is made of two hydrogen atoms and one oxygen atom. We can calculate the **relative molecular mass** for water by adding the relative atomic mass for two hydrogen atoms $(1 + 1)$ and one oxygen atom (16), giving the relative molecular mass of 18. As there is no unit for relative atomic mass, there is no unit for relative molecular mass either.

> ★ | **Worked examples**
>
> 1 Calculate the relative molecular mass of a molecule of oxygen, given that the A_r for $O=16$.
> 2 Natural gas is mainly methane, CH_4. Calculate the relative molecular mass of methane, given that the A_r for $H=1$ and $C=12$.
>
> **Answers**
>
> 1 Oxygen is a diatomic molecule and has the formula O_2, $2 \times O = 2 \times 16 = 32$
> 2 $(1 \times C) + (4 \times H) = (1 \times 12) + (4 \times 1) = 16$

5 What are the units for the relative atomic mass?
6 Calculate the relative molecular mass of ammonia, NH_3.
7 Calculate the relative formula mass of chloromethane, CH_3Cl.
8 Calculate the relative formula mass of methanol, CH_3OH.
9 Suggest the formula for a hydrocarbon that has a relative molecular mass of 30.

▶ The mole

The amount of substance is measured in a unit called the **mole**. One mole of any substance contains 6.02×10^{23} particles. So, 1 mole of oxygen molecules has 6.02×10^{23} oxygen molecules. But, as each oxygen molecule is made of two oxygen atoms (O_2), this means that there are $2 \times 6.02 \times 10^{23} = 12.04 \times 10^{23}$ oxygen atoms in a mole of oxygen molecules.

Key term

Mole The amount of a substance; one mole of any substance contains 6.02×10^{23} particles.

★ Worked example

1 Calculate the mass of one mole of water molecules.
2 Calculate the number of moles of water in 1.5 kg of water.
3 A scientist measured exactly 1.25 moles of water. Calculate the mass of this sample of water.

Answers

1 $M_r = (2 \times H) + (1 \times O) = (2 \times 1) + (1 \times 16) = 18$
  Mass of 1 mole of water $= 18$ g
2 Convert the mass from kg to g (by $\times 1000$). So, there is 1500 g of water molecules in 1.5 kg of water.
  Number of moles (mol) $=$ mass (g)$/M_r$
  Number of moles $= 1500/18 = 83.3$ mol
3 Number of moles (mol) $=$ mass (g)$/M_r$
  Rearrange the equation to get mass as the subject: mass (g) $=$ number of moles (mol) x M_r
  Mass $= 1.25 \times 18 = 22.5$ g

Figure 21.2 One mole of five different elements: carbon (12g), sulfur (32g), iron (56g), copper (63.5g) and magnesium (24g).

✓ Test yourself

10 How many moles of nitrogen molecules are there in 28 g of nitrogen?
11 How many moles of nitrogen atoms are there in 28 g of nitrogen?
12 How many atoms are in 1 mole of helium gas, He(g)?
13 How many atoms are in 1 mole of chlorine gas, Cl_2(g)?
14 Calculate the mass of 1 mole of ammonia (NH_3).
15 Calculate how many moles of carbon are in:
  a) 12 g of graphite
  b) 6 g of diamond
  c) 24 g of buckminsterfullerene
  (You can read more about these different forms of carbon in Chapter 18 Materials for a purpose.)

▶ Sampling

It is important that scientists monitor locations like landfill sites, factories and farms to ensure that pollutants are not leaking into the environment. This monitoring might include checking the water quality in rivers and streams. But, in environmental science, it is not possible to test every drop of water in a river or stream. So, scientists take samples (parts) that are representative (typical) of the whole river.

To understand sampling, think about a bakery. You wouldn't sample every bread roll that is made, because there would be no rolls left to sell. You would sample one bread roll from every batch, testing it to make sure the batch meets the bakery's usual standards. If the sample roll passes the tests, you can assume that all the other bread rolls in the same batch also meet the standard. So, when samples are analysed, it is assumed they have the same properties as the whole substance or all the items in a set of products.

Now consider the processing of drinking water. There are many processing steps to make natural water into the safe drinking water that is pumped to our homes. Samples are taken at particular points in the process and at regular time intervals. The samples are analysed to make sure each processing step is happening correctly. If a sample of drinking water leaving the processing plant is found to be substandard, the testing protocol will help scientists to isolate the point in the production process where the problem is occurring.

Samples should be sealed in specialised clean and dry containers, and labelled with:

- ▶ date and time
- ▶ location (for example part of the river, factory or water processing plant)
- ▶ name of operator taking the sample.

Noting the time and date is important in case additional information, like ambient temperature, is needed to help interpret the results. Including the name of the person who took the sample can be useful if scientists need to ask more questions about when and where the sampling happened in order to interpret the results of a chemical analysis and draw a valid conclusion.

Samples are observed and tested. Some commonly made observations and measurements about samples are:

- ▶ sample mass
- ▶ temperature
- ▶ appearance by visual inspection
- ▶ purity, percentage composition analysis and density by scientific tests.

It is challenging to decide how to sample an environment like a river, as it changes all the time. For example, for some studies it might be appropriate to analyse the water in the centre of the river, at one mile intervals down the whole length of the river. Alternatively, for a different study, scientists might sample the water quality every 10 cm across the width of the river in a single location. Each sample should be representative of the whole at that location. To ensure samples are representative, scientists could take several samples at each sampling point. They can compare results, remove anomalous points and calculate averages.

To provide useful information in a cost-effective and time-effective way, sampling must be properly planned. Poorly planned sampling could lead to harm as scientists might miss the presence of a dangerous substance. For example, a toxic chemical could be concentrated in one area of a landfill site and this might not be apparent if sample locations are too spread out.

⚙ | Specified practical

Titration of a strong acid against a strong base using an indicator

Soluble salts can be made by reacting acids and alkalis together. It is important to add exactly the correct amount of each to ensure the reaction is complete and the maximum yield is obtained.

Procedure

In this experiment 25 cm^3 solution of sodium hydroxide was titrated against 0.1 **mol dm^{-3}** solution of hydrochloric acid. When the indicator changes colour, the neutralisation reaction is complete and has reached its **end point**. The titration can be completed again without the indicator and crystallisation of the resulting solution gives the soluble salt, sodium chloride.

> **Key terms**
>
> **mol dm^{-3}** Moles per cubic decimetre is the unit of concentration. 1 mol dm^{-3} means that there is 1 mole of a substance dissolved in 1 dm^3 (1 litre) of solvent.
> **End point** The point at which the indicator has changed colour in an acid–base titration.

It is also possible to use titrations to accurately determine the concentration of an acid or base. This is useful when stock solutions are prepared, to check that their concentrations are within the tolerance required.

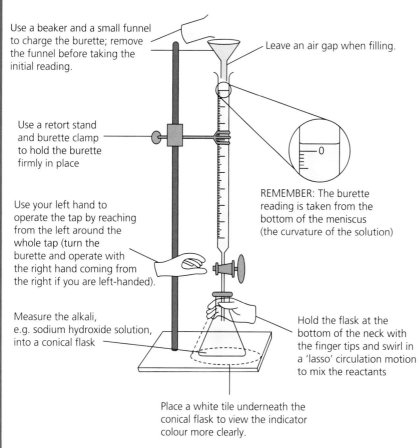

Use a beaker and a small funnel to charge the burette; remove the funnel before taking the initial reading.

Leave an air gap when filling.

Use a retort stand and burette clamp to hold the burette firmly in place

REMEMBER: The burette reading is taken from the bottom of the meniscus (the curvature of the solution)

Use your left hand to operate the tap by reaching from the left around the whole tap (turn the burette and operate with the right hand coming from the right if you are left-handed).

Measure the alkali, e.g. sodium hydroxide solution, into a conical flask

Hold the flask at the bottom of the neck with the finger tips and swirl in a 'lasso' circulation motion to mix the reactants

Place a white tile underneath the conical flask to view the indicator colour more clearly.

Figure 21.3 Method for carrying out a titration

Results

Run	Final volume (cm³)	Initial volume (cm³)	Titre (cm³)
Rough	20.00	0.00	20.00
1	20.50	0.00	20.50
2	25.10	4.50	
3	20.10	10.50	9.40

Analysing the results

1 Write a balanced symbol equation for this reaction.
2 Calculate the missing titre for run 2.
3 Calculate the mean titre.
4 What volume of acid should be added to completely neutralise 25 cm³ solution of sodium hydroxide?

Key term

Titre The total volume of substance added from the burette (titre = final volume reading on the burette − initial volume reading on the burette).

Key term

Data Information for example from observations or measurements.

Analytical techniques

Analytical techniques are standard procedures, so the results are reliable and can be compared between laboratories and operators. In addition to chemical analysis to determine the composition, operators might also use their senses to routinely note the sample's colour, texture and smell.

Data from analysis

Analytical techniques can be classified according to the type of data they produce:

▶ **Qualitative** These data are descriptions of observations. A qualitative analytical technique collects data that are descriptions of observations. An example of a qualitative test would be a flame test where the colour of the flame is observed and noted. This can then be used to draw a conclusion based on the colour of the flame to indicate the metal ion present.

▶ **Quantitative** These data are numerical values to describe a particular variable. A calibrated, precise measuring instrument is used to collect these data. An example of an analytical technique that gives quantitative data is a titration.

▶ **Semi-quantitative** These data include inaccurate numerical values that are generated from uncalibrated measuring instruments or from matching one data point to a reference source. An analytical technique that gives semi-quantitative data is a urine dipstick for measuring glucose concentration. A colour patch on the stick is compared to a reference chart to give an estimated numerical value not a precise measurement (Figure 21.9).

Chromatography

Chromatography can be used to separate and identify substances in a mixture. This technique uses two phases:

▶ A stationary phase to which the sample is added; this is the medium that the mixture travels up or through, for example, paper in paper chromatography.

▶ A mobile phase, the solvent that moves up or through the stationary phase, for example water in the paper chromatography of inks.

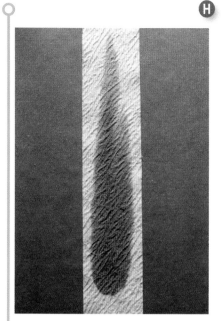

Figure 21.4 Blue dye is more attracted to the mobile phase than the stationary phase and travelled fastest. Whereas the red dye is more attracted to the stationary phase than the mobile phase and travels less quickly

The different parts of the sample mixture are attracted to the stationary and mobile phases by different amounts and this causes them to separate. For example, food colourings are mixes of water soluble dyes. So, water can be used as the mobile phase and absorbent paper can be used as the stationary phase. We find that the different dyes that are used to make the colour, separate and create spots of colour that travel up the paper. The spot that travels the furthest is the most soluble and most attracted to the mobile phase. Whereas, the spots of colour that travel the least are more attracted to the stationary phase than the water. It is the balance of these different attractions that causes each dye to move at a different rate through the stationary phase and produces the chromatogram (Figure 21.4).

The **retention factor**, R_f, can be calculated by:

$$R_f = \frac{\text{distance substance travels}}{\text{distance solvent travels}}$$

R_f is a ratio so doesn't have a unit. It is always a number less than 1 as the solvent travels the furthest in a chromatogram; the bigger the number the more soluble the substance.

In any particular combination of stationary and mobile phases, the R_f value is always the same and can be used to identify the substance by referring to a database of trusted results. The retention factor is affected by temperature (and flow rate for gaseous mobile phase chromatography, see Table 21.1) so these must be controlled for accurate identification.

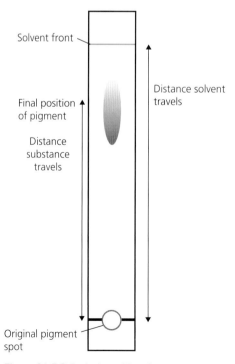

Solvent front

Distance solvent travels

Final position of pigment

Distance substance travels

Original pigment spot

Figure 21.5 Calculation of R_f value

There are four main types of chromatography summarised in Table 21.1.

Table 21.1 Chromatography techniques

Technique	Use	Contexts	How it works	Interpretation
Paper chromatography	Separate and identify components in a mixture (e.g. food dyes/lipstick).	Forensics (e.g. analysis of coloured substances at a crime scene). Food science (e.g. examine food colourings in food).	Stationary phase: absorbent paper. Mobile phase: usually water or ethanol. The different components of the mixture are attracted to the stationary and mobile phases by different amounts, causing them to move up the paper at different rates.	Identify components in a mixture by comparing the chromatogram to known compounds /comparing R_f values.
Thin layer chromatography (TLC)	Separate and identify components in a mixture.	Forensics (e.g. analysis of coloured substances at a crime scene). Food science (e.g. examine food colourings in food). Environmental science (e.g. the analysis of pollutants in the environment).	Stationary phase: thin layer of silica supported on a glass or aluminium plate. Mobile phase: organic solvent, e.g. propanol. The different components of the mixture are attracted to the stationary and mobile phases by different amounts, causing them to move up the plate at different rates. Sometimes the substances cannot be seen with the naked eye and need to be revealed using a chemical reagent (ninhydrin to see amino acids) or UV light (to fluoresce aromatic organic compounds).	Identify components in a mixture by comparing R_f values.
Gas chromatography (GC)	Separate and measure air pollutants. Drugs in hair samples.	Analysis of air quality. Forensic science. Monitoring of athletes for drug use.	Stationary phase: solid support in a column, e.g. high boiling point polymer. Mobile phase: inert gas, e.g. nitrogen or a noble gas. The different components of the mixture are attracted to the stationary and mobile phases by different amounts and emerge from the column at different times. This allows the mixture to be separated and the components to be further analysed.	Use a GLC trace from an analysis to identify an unknown substance using retention time (it is possible for two different substances to have the same retention time).
High performance liquid chromatography (HPLC)	Drugs in blood/ hair samples.	Forensics. Monitoring athletes for drug use.	Stationary phase: fused silica. Mobile phase: organic solvent (liquid). A sample of the solution is injected into the mobile phase. The mobile phase carries the sample through a column that separates the components and each is collected separately.	Use an HPLC trace from an analysis to identify an unknown substance using retention time (it is possible for two different substances to have the same retention time).

Gas phase chromatography can be used to detect the presence of different substances in a mixture. Each peak in the chromatogram shows a different substance in the mixture. The retention time for each peak can be used to identify the substances in turn and the area under the peak indicates their relative concentrations. So, the bigger the area under the peak, the greater the amount of that particular substance in the mixture. The analysis is semi-quantitative because it does not give an absolute concentration but comparable concentrations of the substances in the mixture.

Formulations are special mixtures which have been designed to make a useful product like sun cream or petrol fuel. Chromatograms for formulations are like fingerprints for humans – they are unique. Therefore, if we can compare the overall chromatogram to a database and match it to a registered formulation, we can identify the product.

Figure 21.6 shows the trace produced, where each peak is labelled with a letter to identify it. The values help us to compare the areas under the peaks. One peak corresponds to the mobile phase, and the others are components of the mixture.

1 Suggest the name of the mobile phase.
2 State the number of components in the mixture.
3 Identify which substance has the longest retention time.
4 Write the substances in order from most to least present in the mixture.

Answers

1 The mobile phase is an inert gas, such as nitrogen or any noble gas.
2 The mixture has five peaks and therefore has four substances (one peak is the mobile phase coming out of the machine).
3 Substance E as it has the highest retention time of 20 minutes.
4 The area under the peak (given by the number against each peak) corresponds to the amount of each component. Listed from largest area under the peak to smallest area: E, A, C, D, B

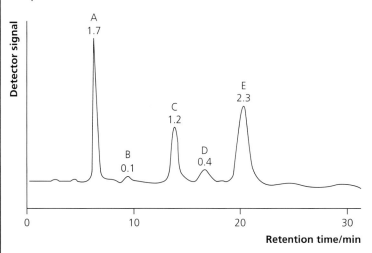

Figure 21.6

Identification of unknown substances using paper chromatography

The ink in coloured pens is formulated from several different dyes mixed together. If a note from a crime scene is taken as evidence, the ink can be analysed using chromatography. The R_f value can then be used to find out the brand of the pen, or even match the ink to that of a pen owned by a suspect.

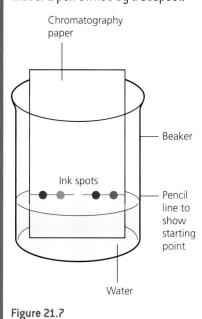

Chromatography paper

Beaker

Ink spots

Pencil line to show starting point

Water

Figure 21.7

Procedure

1 Using pencil, a base line was drawn about 1 cm from the bottom of a piece of chromatography paper.
2 Ink samples were added to the base line. These were spaced equally and labelled in pencil.
3 The prepared chromatography paper was added suspended in a beaker with $0.5\,cm^3$ of solvent, with the bottom end just touching the solvent, allowing the solvent to wick up.
4 The chromatography paper was removed when the solvent line was ahead of all the colours.
5 A pencil mark was made to show the solvent level (solvent front) and the paper was allow to dry.

Analysing the results

1 Why was it important that pencil was used to label the chromatography paper in steps 1 and 2?
2 Why is it important to make a mark of the solvent line in step 4?
3 List the measurements that you would need to take in order to calculate the R_f value for a substance from a chromatogram.
4 Explain how you can use the results from this experiment to determine if any of the inks were the same.

▶ Simple laboratory analysis

Samples of water can be taken from the environment or from crime scenes for analysis. Often, a quick and simple qualitative analysis is undertaken to see if it is worth using a more expensive instrumental analysis for quantitative results. Simple laboratory analysis can be used to determine the presence of different ions, and therefore their compounds, present in a sample. These methods are quick and low cost. Table 21.2 summarises these tests.

Table 21.2 Some simple laboratory analyses and expected results

Test	Outline	Result
Flame test	Put a sample into a blue flame and note the colour.	Sodium, Na^+ – Yellow Potassium, K^+ - Lilac Barium, Ba^{2+} – Green Calcium, Ca^{2+} – Red Copper, Cu^{2+} – Blue/green
Sodium hydroxide precipitate test	Add a few drops of sodium hydroxide solution to the sample and note the colour of any precipitate formed.	Calcium, Ca^{2+} – White Copper, Cu^{2+} – Blue Iron(II), Fe^{2+} – Green Iron (III), Fe^{2+} – Brown Lead, Pb^{2+} – white Chromium (III), Cr^{3+} – Green

Test	Outline	Result
Halide	Add acidified silver nitrate solution to the sample and note the colour of any precipitate formed.	Chloride, Cl^- – White Bromide, Br^- – Cream Iodide, I^- – Yellow
Sulfate	Add barium chloride to the sample and note whether a white precipitate forms.	White precipitate of barium sulfate produced.
Carbonate	Add dilute hydrochloric acid to the sample and note the formation of bubbles of gas.	Effervesces and when gas is tested with limewater this goes cloudy.
Carbon dioxide	Blow the gas through limewater.	Limewater goes from colourless to cloudy.

⚙ Specified practical

Identification of unknown ionic compounds using flame tests and chemical tests for ions

Water samples from rivers and drains are collected regularly by scientists from the Environment Agency to check that factories are not polluting the water. Simple qualitative laboratory techniques can be used to quickly determine the presence of ions. If ions that should not be in the water are detected, more costly quantitative techniques can be used to determine if any safety or legal threshold has been reached and further action needs to be taken.

Procedure

To determine the positive ion (cation):

1 A clean dry nichrome loop was used to transfer a sample of the unknown sample into a blue Bunsen flame and the colour of the flame was noted. If it was not possible to conclude the metal ion was present, step 2 was completed. If a metal ion was found to be present, step 3 was used next.

2 A few drops of sodium hydroxide solution were added to a few drops of the sample solution. The colour of any precipitate formed was noted.

To determine the negative ion (anion):

3 A few drops of acidified silver nitrate were added to a few drops of the sample solution. The colour of any precipitate formed was noted and if no precipitate was formed, step 4 was completed.

4 A few drops of barium chloride solution were added to a few drops of the sample solution. If a white precipitate formed then sulfate ion was present, but if no precipitate formed then step 5 was completed.

5 A few drops of dilute acid were added to a few drops of the sample solution. If effervescence was observed, then a carbonate was present.

Analysing the results

1 Why was a clean dry nichrome wire used in the flame test?

2 Explain why hydrochloric acid should not be used in step 3 to acidify the silver nitrate.

3 If there was a precipitate in step 3, why was it not necessary to continue to step 4?

4 Explain why calcium ions cannot be identified with a positive result in test 2.

5 Suggest the PPE required to complete this investigation.

▶ ## Colorimetry

Nitrate pollution is a long-standing problem in Wales, caused mainly by the run-off from farms of fertilisers, manure and slurry. Regular testing of water is undertaken to monitor the situation.

Nitrate pollution can cause plant overgrowth in waterways, making it difficult for boats to navigate, as well as decreasing biodiversity and particularly causing a fall in the number of fish. In the long term, nitrate pollution can pollute ground water, in turn affecting drinking water quality. High levels of nitrates in drinking water can lead to the potentially fatal 'blue baby syndrome' in which babies' red blood cells do not carry as much oxygen as they should due to toxic levels of nitrates in their body.

The concentration of nitrates in water can be estimated using colorimetry.

Key term

Colorimetry A quantitative technique that can be used to determine the concentration of a solution based on its transmission of light through the sample.

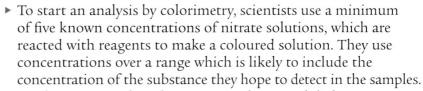

Figure 21.8 Colorimeters are machines that can measure the amount of light going through a sample. The more light that is transmitted, the more dilute the sample. Using a calibration curve from reference samples of known concentration, it is possible to estimate the concentration of an unknown sample

▶ To start an analysis by colorimetry, scientists use a minimum of five known concentrations of nitrate solutions, which are reacted with reagents to make a coloured solution. They use concentrations over a range which is likely to include the concentration of the substance they hope to detect in the samples.

▶ A colorimeter analyses by measuring how much light passes through each measured sample.

▶ The data from the known concentrations is plotted on a calibration curve – concentration (independent variable) goes on the *x*-axis and transmission of light (dependent variable) on the *y*-axis. A line of best fit provides a reference for analysing samples.

▶ The samples contain an unknown concentration of nitrate, which is measured in a similar way to the calibration samples. The transmission of light value is used to determine the concentration of the solution by reading off the line of best fit.

Test strips are a semi-quantitative colorimetry technique. These are quick, cheap and easy to use, but only indicate the concentration of substances rather than giving an accurate value. These strips are made from a combination of coloured reagents that undergo a chemical reaction when they come into contact with the substances that they measure.

▶ DNA profiling

Deoxyribonucleic acid is a natural polymer that is found in the nucleus of cells (page 112). The sequence of bases in this molecule is unique to each individual (unless they are clones). Clones can occur naturally, as in the case of identical twins, or be produced in a laboratory.

Analysing the exact DNA sequence for an individual would be expensive and time consuming, but, by chunking the DNA into smaller sections, it is possible to generate a summary of the DNA called a profile. This broader analysis is usually enough to:

▶ show family relationships
▶ provide evidence for how different species are related through evolution
▶ find if certain genes are present, indicating an increase in the risk of some diseases
▶ match crime scene samples to a suspect.

In **DNA profiling**, DNA is gathered from a biological sample, for example by swabbing the inside of a person's cheek. The steps in the process to form a profile are:

▶ Isolation – separation of the DNA from other tissues.
▶ Fragmentation – Cutting the DNA molecule into smaller chunks using specialist enzymes.
▶ Separation – Using electrophoresis (a special type of chromatography) to separate the fragments to form a pattern.

There are some ethical objections to collecting DNA profiles on a searchable database. A DNA profile is not the same as a complete DNA sequence, but law enforcement agencies can convict on the strength of a DNA profile match. The likelihood of having the same DNA profile as someone else is 100% if you are an identical twin

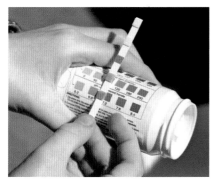

Figure 21.9 To use a test strip, place it in a sample and leave to develop for the time stated in the instructions. Then compare the test strip to the colour charts on the bottle to indicate the concentration of nitrate, lead, chlorine or hydrogen ions (pH) in the solution

Key term

DNA profiling An analytical technique used to determine an individual's characteristics.

and about 1 in 14 million (the same chance as winning the lottery) for the rest of the population. In addition, private companies may gain access to the data. This could mean that people are refused insurance policies due to the risk factors shown in their profile.

▶ Experimental results

Information is collected through experiments and investigations. It is important that scientists understand what the data represents, how it can be displayed and what conclusions could be drawn.

Errors

Accuracy describes how close to the true value a result is. Inaccuracies can be caused by equipment, procedure or human errors. Errors can be classified as:

▶ Systematic – The same error occurs in every piece of data. For example, a top pan balance was not zeroed to $0\,g$ before measurements were taken and displayed a value of $10\,g$. So, each mass that was measured would be $10\,g$ greater than the accurate or true value. These errors are the easiest to correct – you find the inaccuracy and adjust every value accordingly.

▶ Random error – These are errors that are not consistent with each value, making the data difficult to correct. For example, when measuring the resistance of a wire, as the current flows, the wire heats up and this changes the resistance unpredictably. To reduce the effect of random errors, several reliable (similar) results can be used to calculate a mean value.

Displaying data

It is important that the data is displayed in such a way that patterns can be seen and conclusions can be drawn. All data from an investigation should initially be recorded in a suitable results table.

In results tables:

▶ All columns should be clearly labelled with the variable and, if appropriate, the units.
▶ Units should be in the column heading and not after each reading.
▶ The first column should be the independent variable and is completed before the experiment – you choose the values of this variable.
▶ The subsequent columns should be any dependent variables (observations or measurements) that you make during the investigation.
▶ If anomalous results are seen in the table, circle them and do not use them for further analysis.

If it is not possible to see any patterns in numerical data, it can be useful to use visual mathematical methods, for example:

▶ Pie chart – used when the independent variable is categoric data or discontinuous data and the dependent variable is a quantity (%).
▶ Bar chart – used when the independent variable is a category (x-axis) or discontinuous data and the dependent variable is continuous data (y-axis).

Key terms

Anomalous result A result that does not fit in the pattern of the other results.

Categoric data Data that is a category, usually a word, e.g. eye colour.

Continuous data Data that can take any value, e.g. height, handspan, temperature.

Discontinuous data Numerical data that can only have specific values and no intermediate values, e.g. shoe size.

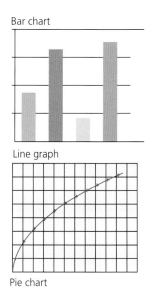

Bar chart

Line graph

Pie chart

Figure 21.10 Data can be displayed in a variety of ways

▶ Line graph – used when both the independent (*x*-axis) and dependent variable (*y*-axis) are quantities. Use crosses not dots to plot points, with the size of the cross indicating your confidence in the accuracy of data; the bigger the cross, the less sure you are of the accuracy of the value. Once the graph is plotted, hold it at arm's length and see if there is a pattern in the data. If there is, draw a line of best fit which shows the pattern – do not draw dot to dot. If there are any points that do not fit the trend, circle them as anomalous points.

How to reach valid conclusions

The aim or objective of an investigation is often to answer a question. Results should be collected to allow you to meet the aim by answering the question. If an investigation actually measures what it set out to measure and gives results that can answer the question, we describe them as valid. Therefore, valid results can be used to support a conclusion which answers the question in the aim.

> ★ **Worked example**
>
> In an experiment a student collected information about the age, height and shoe size of all pupils in class 5F. The aim of their experiment was to find out the average shoe size of children in the class.
>
> 1 Describe how the data could be used to determine the average shoe size of the children in the class.
> 2 Evaluate if the data they collected was valid.
>
> ### Answers
>
> 1 Add all the shoes sizes together and divide by the number of students. Round the number to the nearest half shoe size as shoe sizes only come in whole or half sizes.
> 2 Although the data on the shoe size was valid as it was needed to answer the aim, not all the data is valid. This is because the data about height or age of the student was not needed to answer the aim.

> ⬇ **Chapter summary**
>
>
>
> - Samples for analysis are collected so they are representative of the environment or material.
> - Analytical techniques can be qualitative, semi-quantitative or quantitative.
> - The mole is a unit to measure the amount of substance.
> - One mole of any substance has 6.02×10^{23} particles of that substance in it and the same numerical mass value as the relative atomic mass for single atoms or relative molecular mass for molecules but measured in grams.
> - Chromatography is used to separate substances in a mixture to identify them using the retention time R_f.
> - Chromatography techniques include paper, TLC, HPLC and GLC; all have a mobile phase and a stationary phase.
>
> - Colorimetry is a qualitative technique which measures the transmittance of light through a solution to determine the concentration.
> - Coloured test strips are a semi-quantitative colorimetry analysis.
> - DNA is unique to individuals (except identical twins) and so genetic profiling can be used to identify criminals, family relationships and genetic diseases.
> - Data must be recorded in a suitable way in order that patterns can be seen and valid conclusions drawn.
> - Errors are inaccuracies in data and they can be both random and systematic errors.

► Practice exam questions

1 An environmental health officer took samples of soft drinks at a café to check they were correctly labelled and safe to drink. Some tests were completed at the café and samples were also taken to the laboratory to be tested.

   a) Describe how the environmental health officer could measure the pH of the samples in the café. [2]

   b) The Environmental Health Officer used a test strip to measure the concentration of nitrates in the drink. Classify this test as qualitative, semi-quantitative or quantitative. [1]

   c) The scientists in the laboratory analysed the drinks using TLC. The figure shows the chromatogram.

   Calculate the R_f value for the red colouring, P. [3]

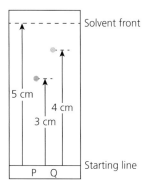

Figure 21.11

2 Some labels fell off three white powders in a chemical store cupboard. An analytical chemist was asked to determine what the substances were. The chemist used laboratory tests to determine the substances present:

   a) Substance A gave a yellow flame test and had effervescence when dilute hydrochloric acid was added. Give the name of this substance. [2]

   b) Substance B gave a white precipitate with sodium hydroxide solution and also a white precipitate with acidified silver nitrate. It also gave a red flame test. Give the formula of the substance. [2]

   c) Substance C made a green precipitate with sodium hydroxide solution and a white precipitate with barium chloride. Justify why this substance is not iron(III) sulfate. [3]

22 Controlling chemical reactions

▶ Reactions, reactants and products

Chemical reactions are changes that occur when a new substance is made. It is important to control the speed (or rate) of chemical reactions so that they are safe for us and for the environment, and so they make profitable amounts of products efficiently.

Raw materials contain the reactants that undergo a chemical reaction to make desired products. Sometimes energy is needed to start the chemical change. Reaction products have to be separated and purified before they can be used, especially when making pharmaceuticals.

▶ Energy

Atoms in substances are held together by bonds. These bonds are stores of chemical energy. Different substances contain different amounts of stored chemical energy, depending on the number and nature of their bonds.

Chemical reactions can be:

▶ **exothermic** – energy is released, causing a temperature increase in the surroundings
▶ **endothermic** – energy is taken in, causing a temperature decrease in the surroundings.

Reaction profiles

A model is a simplification that helps us to understand a process. This is a model of what happens in a chemical reaction:

1 Energy is needed to break the chemical bonds between the atoms in the reactants to make individual atoms. This energy is called the **activation energy**.
2 Individual atoms rearrange to make the structure of the products.
3 New bonds form to make the products and energy is released.

Reaction profiles or energy level diagrams can show the energy changes that happen in a chemical reaction. The *x*-axis represents the time over which the chemical reaction happens. The *y*-axis shows the relative energy levels of the reactants and products, but does not give an absolute value of the energy in the substances.

An input of energy – the activation energy – is needed to start any reaction. This is the energy needed to break the bonds in the reactants to form separate atoms which can then rearrange. This intermediate step with separated atoms is at a higher energy level

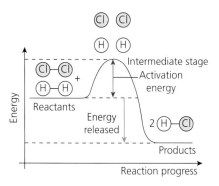

Figure 22.1 An exothermic reaction. The energy needed to break the bonds in the reactants is less than the energy that is released when the products are formed

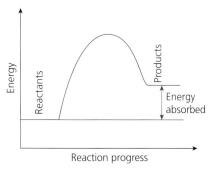

Figure 22.2 An endothermic reaction. The energy needed to break the bonds in the reactants is more than the energy that is released when the products are formed

than that of the reactants. If energy is released to the surroundings in an exothermic reaction, the energy level of the products is lower than the energy level of the reactants. If the reaction is endothermic, the energy level of the products is higher than the reactants. The difference between the starting and final energy levels is the energy that is released in an exothermic reaction or taken in by an endothermic reaction.

Let's consider the reaction to make hydrogen chloride gas from its elements. The equations for this reaction are:

hydrogen + chlorine → hydrogen chloride
$H_2(g) + Cl_2(g) \rightarrow 2HCl(g)$

In this reaction, the reactants have a certain energy level, the activation energy is applied and breaks all the bonds in the elements. This makes separate atoms as an intermediate stage. This is an unstable state and has high energy. The atoms rearrange to make the products and energy is released as the new bonds form. Overall, the energy stored in the products is less than the energy stored in the reactants, so energy is released and this reaction is exothermic.

> ✓ **Test yourself**
>
> 1 What happens in a chemical reaction?
> 2 Classify the following reactions as exothermic or endothermic:
>   a) Combustion of a wax candle.
>   b) Dissolving of ammonium chloride into water in a cold pack.
>   c) Explosion of a firework.
> 3 What is the name of the energy needed to break reactant bonds?

▶ Rate of reaction

Chemical reactions can happen at different speeds, for example:

- ▶ rusting is slow
- ▶ cooking is fast
- ▶ fireworks are very fast (explosive).

The speed or **rate of a chemical reaction** is measured by a change in the amount of a reactant or a change in the amount of a product, **in a given time**. This value can be calculated from experimental measurements.

Measuring the rate of reaction

Understanding rate of reaction is important in the chemical industry. Engineers can predict how long it will take to make a product and know which factors to change to improve efficiency, maximise productivity and minimise costs. They can also control reactions by modifying the rates based on availability of raw materials, demand for the end product and safety considerations.

Scientists look at the balanced symbol equation, consider the properties of each reactant and product, and think about the best way to monitor the reaction. Different methods are evaluated before a decision is made on how the reaction will be monitored.

To monitor the rate of reaction you need to be able to determine how much of a reactant is used in a given time or how much product is made in a given time.

For example:

▶ pH can be monitored with a pH probe, indicating the concentration of $H^+(aq)$ in solution and therefore the concentration of the acid. If acid is a reactant, the pH will increase over the course of the reaction. If acid is a product, the pH will decrease over time.
▶ In some chemical reactions a colour change is observed. It is possible to use colorimetry (see page 184) to measure the colour intensity and estimate the concentration of coloured species in solution.
▶ TLC (see page 181) can be used to determine the progress of an organic chemistry reaction. Samples are taken from the reaction mixture at time intervals to determine if any reactants are present and to show when the reaction is complete.

There are three common methods for monitoring the rate of a reaction in the laboratory. To decide which method to use, think about the characteristics of the reaction and whether it is easiest to monitor disappearance of a reactant or the production of a product:

▶ Volume of gas produced – when you collect a gas, you need well-fitting equipment so that none is lost to the atmosphere. Using a gas syringe is more accurate than displacing water from an inverted measuring cylinder, but a gas syringe holds a smaller volume of gas (Figure 22.3). The volume of gas collected is noted at equally spaced time intervals. These data can then be plotted on a graph (gas volume versus time) and used to determine the mean rate of reaction or, by using the gradient of part of the graph, the rate of reaction at any point during the experiment.

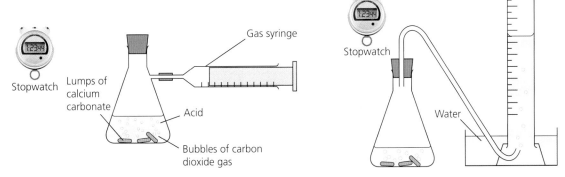

Figure 22.3 Measuring the rate of reaction when calcium carbonate reacts with nitric acid in acid rain to make calcium nitrate solution, water and carbon dioxide gas. As this reaction releases carbon dioxide gas, we can collect and measure the volume of gas produced in a certain amount of time. The volume of gas is measured by gas syringe or by displacement of water from an inverted measuring cylinder

▶ Mass change – this method can only be used in open systems where substances can enter or leave the reaction vessel (Figure 22.4 and 22.5). Only small changes in mass are observed in laboratory-scale reactions, so very precise and sensitive balances are needed to observe these reactions. Of course, the overall mass of a reaction system is constant, but mass can appear to increase if a gas from the air is a reactant that is incorporated into a product, or mass can appear to decrease if a gas is produced and lost to the atmosphere. The mass of the reaction vessel should be recorded at equally spaced time intervals, where the time is such that mass changes can be observed. For some fast rates of reaction (such as magnesium + acid) the time interval might be every 10 seconds, but for slow reactions like rusting the time interval would be much longer.

Figure 22.4 Measuring the rate of reaction when iron rusts using a top-pan balance. As this reaction is an oxidation reaction, the mass appears to increase due to the gain of oxygen from the air

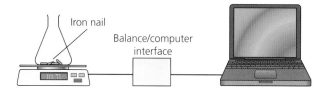

Figure 22.5 When measuring the loss or gain in mass during a very slow reaction, it is possible to use a data logger to conveniently record the measurements over a long period of time

▶ **Turbidity** – in this method, we monitor formation of an insoluble precipitate, which makes the reaction solution cloudy and obscures a cross marked under the reaction vessel (Figure 22.6 and 22.7). A series of experiments is completed, changing one variable in increments. The observer times how long it takes for the cross to disappear, which is assumed to be at the point at which the reaction is complete. The different runs of the experiment can be compared, with the longer times indicating slower reaction rates.

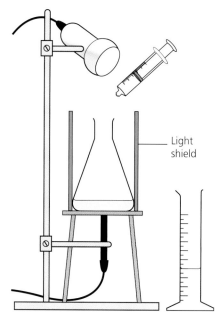

Figure 22.7 It is possible to monitor the transmission of light through a reaction mixture using a light detector and data logger. This allows for constant monitoring as the precipitate forms and is more reliable as it reduces the human error – in Figure 22.6 observers may see the cross disappear at different times based on their own eyesight and perception

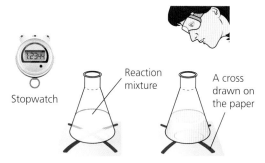

Figure 22.6 When sodium thiosulfate solution ($Na_2SO_2O_3$) reacts with dilute hydrochloric acid, a yellow precipitate of sulfur turns the solution cloudy. You can measure the rate of reaction by timing how long it takes for a cross on paper under the reaction mixture to be obscured. This is called the disappearing cross method

Reaction rate graphs

Data collected from reaction rate experiments are written up in tables, but it can be difficult to see patterns and draw conclusions. So, often the data are plotted on a graph, with time on the x-axis and the dependent variable that was used to monitor the rate of reaction (for example, mass or volume) on the y-axis.

The line of best fit can be drawn, showing the trend in the data. If the whole reaction is monitored, the rate of reaction graph will be a curve that flattens off at the end. But in many reactions, only the first part of the reaction is monitored, so the graph is a straight line which shows a proportional or directly proportional relationship. The classic rate of reaction curve would be seen with more data points later in the reaction.

The **gradient** of the line of a reaction rate graph indicates the rate of reaction: the steeper the gradient, the faster the rate of reaction. When the reaction is complete, time continues but the y-axis value doesn't change, so the gradient is 0 and the rate of reaction is also 0.

The gradient of a straight line can be calculated by choosing two points on the line of best fit with values that are easy to read off the scales. As a general rule, try to use points that fall on the grid lines of the graph paper. Then use the formula:

$$\text{gradient} = \frac{\text{change in } y}{\text{change in } x}$$

To find the gradient of a reaction rate graph with a curve (not a straight line), a **tangent** must be drawn. The gradient of the tangent tells you the rate of reaction at that specific point on the curve (Figure 22.8).

Key terms

Gradient The size of the slope of a straight line on a graph.

Tangent The straight line that best represents a small section of a curved trend line on a graph.

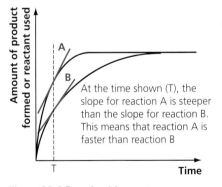

At the time shown (T), the slope for reaction A is steeper than the slope for reaction B. This means that reaction A is faster than reaction B

Figure 22.8 Reaction A has a steeper gradient and therefore a faster reaction rate compared to B. Both lines level out at the same value on the y-axis, showing that the same amount of product was formed

A student collected the gas evolved from a reaction between calcium carbonate and dilute hydrochloric acid for 1 minute. The table below shows her results.

Results

Time (s)	Volume (cm³)
10	10
20	19
30	31
40	40
50	51
60	31

1 Plot a rate of reaction graph of volume against time for these data.
2 Determine which is the anomalous result and suggest a cause.
3 Calculate the rate of reaction for this reaction.
4 Evaluate whether this reaction was complete.

★ | Worked example

A student investigated the effect of changing concentration of hydrochloric acid on the rate of reaction with magnesium. The student used three different concentrations of hydrochloric acid and measured the volume of gas that was produced over 15 minutes. They drew a graph of their results.

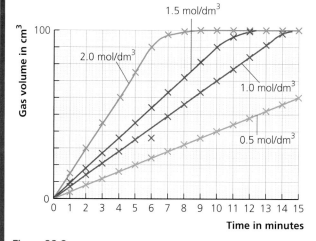

Figure 22.9

1 Write a balanced symbol equation, including state symbols, for this reaction.
2 State the independent variable.
3 State the unit of the dependent variable.
4 Classify time as an independent, dependent or control variable and justify your answer.
5 Describe the trend shown in the lowest concentration data.
6 Explain how you can use the graph to determine how concentration affects rate of reaction.

Answers

1 $HCl(aq) + Mg(s) \rightarrow MgCl_2(aq) + H_2(g)$
2 The independent variable is the one that is being changed, which is the concentration of hydrochloric acid.
3 Time is a control variable as it is the same for each concentration that is being investigated.
4 The trend is directly proportional. The trend line starts at the origin and if the time is doubled, then the volume also doubles.
5 The greater the volume of gas that is collected in the same time, the faster the rate of reaction. The rate of reaction can be calculated from the gradient of the graph. Looking at the graph, the steeper the trend line the faster the rate of reaction.

(H)

► Collision theory

Chemical reactions are difficult to understand as they involve changes at the atomic level. A model can help us imagine how chemical reactions happen and help us predict the effects of changing conditions.

Collision theory is a model, or simplified version, of what happens during a chemical reaction. This theory supposes that reactants are particles that must collide for a chemical reaction to happen. Most collisions do not actually lead to product formation, so the theory states that the collisions must be of *high enough energy* for a reaction to happen.

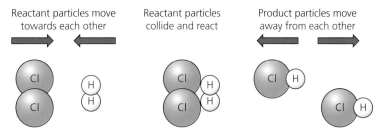

Reactant particles move towards each other

Reactant particles collide and react

Product particles move away from each other

Figure 22.10 Collision theory states that reactant particles must collide with enough energy for a chemical reaction to take place

At the start of a chemical reaction, only reactant particles are present, so:

▸ there are no product particles
▸ the chance of a successful collision between reactant particles is high
▸ rate of reaction is highest.

As the reaction progresses, the concentration of reactants decreases and the concentration of products increases. This causes the rate of reaction to slow. Towards the end of the reaction, there are very few reactant particles left so the number of successful collisions is low and rate of reaction is very slow. When there are no reactant particles left, there are no successful collisions and the rate of reaction is zero; the reaction is complete.

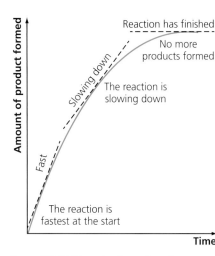

Figure 22.11 The rate of reaction changes as the reactant particles react and become the product particles

Lower temperature

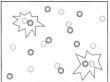

Lots of collisions don't produce a reaction

Higher temperature

Higher number of useful collisions

Figure 22.12 At lower temperatures there are still collisions but the particles are too low energy for these to be successful. At higher temperatures there are more collisions overall and so more successful collisions, leading to a higher rate of reaction

Key terms

Thermal runaway Can occur in an exothermic reaction, when increasing the temperature increases the rate of reaction, which then increases the temperature further.

Reaction pathway A series of different steps that happen in a chemical reaction to get from the reactants to the products.

The more collisions there are per second, the greater the number of successful (reacting) collisions and the faster the rate of reaction. There are several factors that affect the rate of reaction as they affect the number of successful collisions in a given time:

▶ temperature
▶ concentration
▶ pressure
▶ surface area
▶ catalysts.

Temperature

Increasing the temperature increases the rate of reaction because:

▶ the reactant particles move faster, so there are more collisions in a given time.
▶ the reactant particles have more kinetic energy, so it is more likely that the activation energy will be supplied and any collisions will be successful.

As the temperature increases, the rate of an exothermic reaction increases, thereby providing further heating and causing an even faster rate. This is a positive feedback loop in which a change causes a further similar change. Chemical engineers must carefully control exothermic reactions because of the potential for thermal runaway, which could lead to an explosion. They can design processes to limit temperature rises by using:

▶ batch production, which restricts the amount of reactants used and so limits the amount of energy released. The production is more easily controlled.
▶ alternative reaction pathways, which may take more time but are more easily controlled. For example, in the contact process to make sulfuric acid, water cannot be added directly to sulfur dioxide as the reaction is too exothermic. However, the sulfur dioxide can be reacted with oleum (concentrated sulfuric acid) and is then diluted to make sulfuric acid. This reaction pathway releases energy more slowly but takes more time and requires an additional raw material, oleum.
▶ specialist reaction vessels with automated cooling systems.
▶ catalysts, which lower the activation energy of the reaction and so limit the amount of heating needed.

Thermal runaway happened in 1947 at Texas City, Galveston Bay, in a fire that started on board a ship docked in port. Sadly, the SS Grandcamp was carrying about 2100 tonnes of ammonium nitrate, which can be used as a fertiliser, but is also explosive. An explosion started a chain reaction and the high temperatures caused nearby ships and oil-storage facilities to catch fire, leading to the deaths of 581 people.

A more recent example is the Bhopal disaster in India, which happened overnight in early December 1984. A pesticide plant accidently leaked a toxic gas, methyl isocyanate, exposing over 500 000 people. The gas leak happened when water entered a faulty storage tank, causing a runaway exothermic reaction.

Concentration and pressure

The concentration of a solution is a measure of the amount of solute that is dissolved in a volume of solvent. The more concentrated the solution, the more solute there is in a given volume of solution.

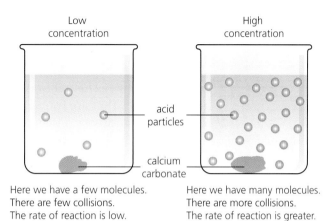

Here we have a few molecules.
There are few collisions.
The rate of reaction is low.

Here we have many molecules.
There are more collisions.
The rate of reaction is greater.

Figure 22.13 In dilute solutions there are less reactant particles in a given volume than in concentrated solutions. This leads to a lower number of successful collisions in a given time and therefore a lower rate of reaction

Increasing the pressure of the gas gives the same number of gas particles in a smaller volume of space, effectively increasing the concentration of gas particles.

Increasing the concentration of the reactants in solution or the pressure of a gas increases the rate of reaction because there are more reactant particles per volume of solution and so there are more collisions in a given time. Increasing the number of collisions overall increases the number of successful collisions.

Surface area

Increasing the surface area of a solid reactant increases the rate of reaction because more of the solid reactant particles are exposed and so able to collide with the other reactant.

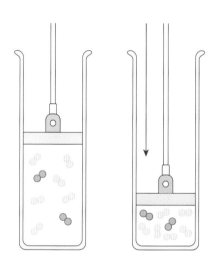

Figure 22.14 Increasing the pressure by depressing the plunger on a syringe decreases the volume and so the molecules have less space to move. They are more likely to collide and the reaction rate increases

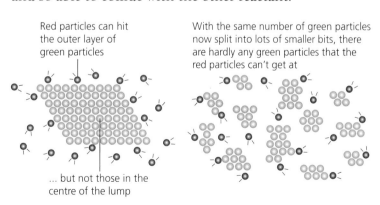

Red particles can hit the outer layer of green particles

With the same number of green particles now split into lots of smaller bits, there are hardly any green particles that the red particles can't get at

... but not those in the centre of the lump

Figure 22.15 In lumps of solid material there are less reactant particles on the surface available for collision, compared to the same mass of a powder. This leads to a lower number of successful collisions in a given time and therefore a lower rate of reaction

So, the rate of reaction is faster for a powder than for a larger lump of the same material.

Key term

Catalyst Chemical that increases the rate of a reaction by reducing the activation energy without itself undergoing any permanent chemical change.

A **catalyst** increases the rate of a reaction. Each catalyst is specific to a particular chemical reaction but is never used up in that reaction. For example, iron is a catalyst in the Haber process to make ammonia for fertiliser.

Catalysts often provide a surface for the reaction to take place and this lowers the activation energy of the reaction. There is no change in the number of collisions, but more will be successful in the same amount of time, so the rate of reaction increases.

Catalysts can benefit the environment. For example, catalytic converters change the pollutant gases from the car exhausts into harmless gases that are found naturally in air. Catalysts are also very important in industry. They save time and money by increasing the efficiency of chemical reactions. By lowering the activation energy of reactions, catalysts decrease the heating required to bring about chemical change, thereby limiting fossil fuel consumption and the production of greenhouse gases. Research focuses on finding new catalysts that are cheaper to make and easier to use but have the same or enhanced effects as existing catalysts.

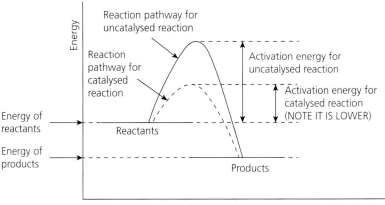

Figure 22.16 Catalysts are substances that provide an alternative reaction pathway with lower activation energy. This means that lower energy collisions can now be successful

Catalysts are very important in industry as they:

▶ reduce the operating temperatures of reactions, reducing energy costs of providing the high temperatures for the activation energy to start the reaction. As catalysts can be reused many times they are an economical choice compared to running at high temperatures.
▶ increase the actual yield. As catalysts are specific, they ensure the desired product is made rather than unwanted side products.
▶ reduce the time it takes to make the product, making the process quicker and more profitable.
▶ preserve raw materials, as less waste is generated.

As catalysts do not get consumed in reactions they can be used over and over again. They do need to be cleaned from time to time to ensure that they work at their best.

A catalyst often provides a surface for the reaction to happen on, thus lowering the activation energy. So, in industry, catalysts are often used as powders to increase their surface area and meaning that less catalyst is required. This is desirable as many catalysts are relatively expensive transition metals.

Test yourself

7 What changes can be made to increase the rate of a reaction?
8 What is collision theory?
9 What is needed for a successful collision?
10 Explain how these factors affect the number of successful collisions in a given time:
   a) increase temperature
   b) increase concentration of reactants
   c) increase pressure of gas reactants
   d) adding a catalyst

Specified practical

Investigation of the factors affecting the rate of reaction between dilute hydrochloric acid and sodium thiosulfate

Sodium thiosulfate reacts with hydrochloric acid to form a gas and an insoluble precipitate:

$$Na_2S_2O_3(aq) + 2HCl(aq) \rightarrow 2NaCl(aq) + S(s) + SO_2(g) + H_2O(l)$$

The reaction can be monitored by measuring the loss of mass as the gas product escapes to the surroundings or by measuring the volume of gas collected. These methods give data points throughout the time the reaction is monitored. When a graph of mass or volume against time is plotted, the gradient at any point allows calculation of the rate of reaction at that time.

As the reaction produces a precipitate giving a cloudy (turbid) appearance, it can also be monitored using the disappearing cross method. For this method, a time point is recorded: the time taken for a cross drawn under the reaction vessel to disappear as the solution becomes turbid. This method is limited in that it can only give an *average* rate of reaction for each individual experimental run. The time it takes for the cross to disappear indicates that the reaction is complete. It takes less time for the cross to be obscured at a faster rate of reaction. This method is suitable for comparing reaction rates with different concentrations of reactants or reactants at different temperatures.

A student decided to investigate how the temperature affected the rate of reaction and used the disappearing cross method to monitor the reaction.

Sodium thiosulfate solution

Sodium thiosulfate and dilute acid

Figure 22.17

Results

Temperature (°C)	Time for the cross to disappear (s)
20	79
30	58
40	31
50	20
60	8

Analysing the results

1 Name the gas product of this reaction.
2 Name the precipitate from this reaction.
 3 Justify the use of the disappearing cross method to determine the effect of changing the concentration of the hydrochloric acid.
4 Why is eye protection worn?
5 What is the independent variable?
6 What is the dependent variable?
7 What are the control variables?
8 Plot these data on a graph (with temperature on the *x*-axis, and time for the cross to disappear on the *y*-axis) and draw a suitable best-fit line.
9 Describe the pattern in the data.
10 A student suggests that if the temperature is doubled, then the rate of reaction doubles. Do you agree? Explain your answer.

Chapter summary

- In chemical reactions a new substance is made.
- Chemical reactions that give out energy are described as exothermic and cause a temperature increase in the surroundings.
- Chemical reactions that take in energy are described as endothermic and cause a temperature decrease in the surroundings.
- Sometimes the rate of exothermic reactions can accelerate as temperature rises and this can lead to thermal runaway and explosions.
- Rate of reaction is a measure of the speed at which a reaction takes place. This can be monitored by measuring the mass changes, collection of gas or time it takes for a precipitate to form.
- Rate of reaction can be increased by increasing temperature, increasing the surface area of solid reactants, increasing the pressure of gaseous reactants, increasing the concentration of solution reactants or by adding a suitable catalyst.
- Collision theory can help make predictions on how changing conditions will affect the rate of reaction. Reactions can only happen if reactant particles collide with enough energy.
- The minimum amount of energy for a reaction to happen is called the activation energy. Catalysts increase the rate of reaction by providing an alternative reaction pathway with a lower activation energy, without being used up in the reaction themselves.
- Catalysts are specific to reactions and reduce the cost of completing industrial processes involving chemical reactions by reducing the operating temperature, increasing the yield and preserving raw materials.

23 Controlling nuclear reactions

▶ Nuclear power

Do we want to build more nuclear power stations? On the one hand, nuclear power goes a great way towards generating large quantities of 'on-demand', carbon-neutral electricity, but, on the other hand, the events of 1 March 2011, when an 8.9 magnitude earthquake 400 km north-east of Tokyo triggered a 14-metre tsunami which hit the shore at the Fukushima nuclear power plant, have made the decision to build new reactors in the UK a much more difficult proposition. Do the risks associated with nuclear power outweigh the risks associated with burning fossil fuels, the emission of greenhouse gases, and global warming?

How does nuclear power work?

Atoms are made up of very small positively charged nuclei with electrons orbiting the nucleus. The nuclei are made up of nucleons (protons and neutrons) and the $_Z^A X$ notation is used to describe the nucleus where A is the nucleon number (or mass number), Z is the proton number (or the atomic number) and X is the atomic symbol.

Nuclear-reactor fuel is uranium, which consists of two main isotopes – uranium-238 ($_{92}^{238}U$) and uranium-235 ($_{92}^{235}U$). As they are both forms of uranium, they have the same number of protons ($Z = 92$), but they have different numbers of neutrons. In uranium-238, $A = 238$ and $Z = 92$, so the number of neutrons is $238 - 92 = 146$; and in uranium-235, $A = 235$ and $Z = 92$, so the number of neutrons is $235 - 92 = 143$. Uranium-235 is the isotope used inside nuclear reactors because it undergoes nuclear fission.

▶ Nuclear fission

All current nuclear reactors use the process of nuclear fission to produce their primary source of energy – the word 'fission' means 'breaking up'. Uranium-235, $_{92}^{235}U$, is naturally radioactive and decays via alpha decay ($_2^4He$) into thorium–231, $_{90}^{231}Th$. This is summarised by the nuclear equation:

$$_{92}^{235}U \rightarrow {}_{90}^{231}Th + {}_2^4He$$

When writing nuclear equations using the $_Z^A X$ notation, a conservation law applies; all the values of A (the mass number) on the left-hand side of the equation must add up and equal the addition of all the values of A on the right-hand side. The same law applies to the values of Z (the atomic number).

Key terms

Nuclear fission When a large unstable nucleus breaks up spontaneously, or on impact with a neutron, releasing energy.

Chain reaction When one nuclear fission produces several neutrons which go on to produce further fissions, which also go on to produce further fissions, and so on.

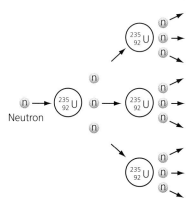

Figure 23.1 A chain reaction in uranium-235

Inside a nuclear reactor, U-235 nuclei can be broken up into large 'daughter' nuclei (rather than alpha decay), if they are bombarded by slow-moving neutrons – this process is called **nuclear fission**. A typical nuclear equation for this process is shown by:

$$^{235}_{92}U + ^{1}_{0}n \rightarrow ^{141}_{56}Ba + ^{92}_{36}Kr + 3^{1}_{0}n$$

During the fission, more free neutrons are produced, which themselves, in turn, can produce the fission of other U-235 nuclei, and so on, starting a process called a **chain reaction** (Figure 23.1).

An uncontrolled nuclear chain reaction can lead to an explosion which releases huge amounts of energy in a very short period of time. This mechanism is used in nuclear weapons. Nuclear reactors however, use control mechanisms to limit the speed of the chain reaction and release the energy much more slowly.

✔ Test yourself

1. 1 atom of U-235 has a mass of 3.9×10^{-25} kg. How many atoms of U-235 are there in 1 kg? If each atom's nucleus can emit 3.2×10^{-11} J of energy, how much energy could 1 kg of U-235 produce?

2. 1 kg of U-235 could produce about 83 TJ (83×10^{12} J) of energy. By comparison, 1 kg of best coal could produce 35 MJ (35×10^6 J). How much coal would you have to burn to get the same amount of energy as 1 kg of uranium-235?

3. Are there any other considerations when comparing coal and uranium as fuels for generating electricity?

4. Complete the following nuclear equations for fissions that occur inside a nuclear reactor:

    a) $^{235}_{92}U + ^{1}_{0}n \rightarrow ^{...}_{53}I + ^{95}_{39}Y + 3^{1}_{0}n$

    b) $^{235}_{92}U + ^{1}_{0}n \rightarrow ^{144}_{56}Ba + ^{90}_{36}Kr + ...^{1}_{0}n$

How does fission work?

The fission of U-235 breaks the nucleus up into two daughter nuclei, one with a nucleon number of about 137 and another with a nucleon number of about 95. On average, three neutrons are also produced, but this can be up to five or as low as one. The precise decay that any one U-235 nucleus undergoes depends on many factors, including the speed of the incoming fission neutron. We can represent one common decay by the nuclear equation:

$$^{235}_{92}U + ^{1}_{0}n \rightarrow ^{144}_{56}Ba + ^{89}_{36}Kr + 3^{1}_{0}n$$

In this example, the U-235 nucleus is impacted by a slow-moving neutron. The two daughter nuclei that are formed are barium-144 and krypton-89, and three other neutrons are produced that, when slowed down by a moderator inside a reactor, can go on to form three further fission events (Figure 23.2).

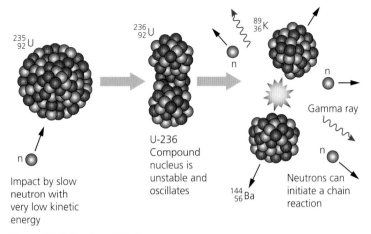

Fission yields fragments of intermediate mass, an average of 2.4 neutrons, and high kinetic energy

$^{235}_{92}$U

$^{236}_{92}$U

$^{89}_{36}$K

n

Gamma ray

$^{144}_{56}$Ba

n

Impact by slow neutron with very low kinetic energy

U-236 Compound nucleus is unstable and oscillates

Neutrons can initiate a chain reaction

Figure 23.2 Uranium-235 decay

Many of the daughter fission products are also radioactive and they decay with a large range of half-lives, from iodine-129, which has a half-life of 15.7 million years, down to europium-155, with a half-life of 4.76 years. Nuclear fuel rods remain radioactive for a very long time and need to be stored very securely.

> ### ✔ Test yourself
>
> 5 Use a Periodic Table to write nuclear equations to summarise the following fission reactions inside a nuclear fuel rod, occurring from the fission of uranium-235 from one neutron. The equations are of the form:
>
>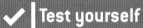
> $$^{235}_{92}\text{U} + ^{1}_{0}\text{n} \rightarrow \ldots + \ldots + \ldots ^{1}_{0}\text{n}$$
>
> The fission products are:
> a) xenon-140, strontium-94 and two neutrons
> b) rubidium-90, caesium-144 and two neutrons
> c) lanthanum-146, bromine-87 and three neutrons.

▶ Reactor engineering

The nuclear fission process is only possible if the neutrons that are released from the fission of uranium-235 are moving slowly enough. If the neutrons are moving too fast, they will not cause fission. The slow-moving neutrons are called **thermal neutrons**. In order to slow down the fast-moving neutrons produced by the nuclear fission process, the fuel rods in the reactor are surrounded by a material called a **moderator**. Most nuclear reactors use water as the moderator (these are called pressurised water reactors or PWRs (Figure 23.3)), and others use graphite rods (graphite is a form of carbon). The advantage of using water as a moderator is that it can also act as the coolant and the mechanism of heat transfer for the reactor. The hot water is used to make steam which turns a turbine; this drives a generator that produces electricity. In the event of loss of coolant, the nuclear chain reaction stops (the neutrons are moving too fast), but the reactor overheats; this is one of the things that happened at the Fukushima nuclear plant following the tsunami of March 2011.

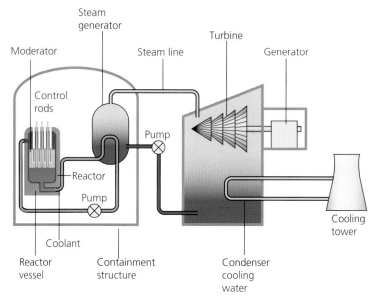

Figure 23.3 Pressurised water reactor

The whole nuclear-fission process can be completely stopped, speeded up or slowed down by controlling the number of thermal neutrons inside the fuel rods. In a nuclear reactor, this is achieved by inserting neutron-absorbing rods, called **control rods**, into the spaces between the fuel rods. Materials such as boron, cadmium and hafnium are commonly used to make control rods. All modern reactors have 'fail-safe' mechanisms built into their systems so that, if a fault occurs, the control rods automatically drop down into the reactor, shutting off the chain reaction. Moving the control rods down into the reactor slows the reaction down (or stops it completely) by absorbing more of the thermal neutrons; moving the control rods up speeds up the reaction as fewer thermal neutrons are absorbed.

The safety of nuclear reactors is further improved by encasing the reactor inside a strong steel pressure vessel, and then surrounding the whole reactor vessel inside a containment structure made of concrete. The combined effect of the pressure vessel and the containment structure is to prevent any radioactivity from escaping.

Half-life

The **activity** of a radioactive sample (measured in becquerels, Bq) is the number of radioactive decays per second. The naturally occurring background activity of a school laboratory is about 0.5 Bq, and a standard school radioactive source has an activity of about 150 kBq (150 000 Bq); a spent nuclear fuel rod, however, would have an activity of 46 000 TBq (46 000 000 000 000 000 Bq!). The decay time of radioactive atoms is compared using a measurement called **half-life**. The half-life of a radioactive substance is the time it takes for the activity of a sample to halve.

Table 23.1 Half-lives of some radioactive atoms

Radioactive atom	Half-life (millions of years)
Technetium-99	0.211
Selenium-79	0.327
Zirconium-93	1.53
Caesium-135	2.3
Palladium-107	6.5
Iodine-129	15.7

The radioactive atoms in Table 23.1 are all found in spent nuclear fuel and have very long half-lives. The safe storage of spent nuclear waste is a huge issue for human beings. The storage facilities will need to be kept safe for HUNDREDS of MILLIONS of years.

★ | Worked example

A small sample of zirconium-93 inside a glass block has an initial activity of 1200 kBq. Use the information in Table 23.1 to determine the number of years and activity of the sample after:

a) 1 half-life
b) 3 half-lives
c) 5 half-lives.

Answer

a) 1 half-life = 1.53 million years; activity $= \dfrac{1200\,\text{kBq}}{2} = 600\,\text{kBq}$

b) 3 half-lives = 3 × 1.53 million years = 4.59 million years; activity
  = 1200 kBq ÷ 2 ÷ 2 ÷ 2 = 150 kBq

c) 5 half-lives = 5 × 1.53 million years = 7.65 million years; activity
  = 1200 kBq ÷ 2 ÷ 2 ÷ 2 ÷ 2 ÷ 2 = 37.5 kBq

▶ Nuclear disasters

There have been three major nuclear power plant disasters:

▶ **Three Mile Island, Pennsylvania, USA, 1979** – caused by a combination of mechanical failure and human error, resulting in the partial melt-down of the reactor core and the release of radioactive coolant water into the surrounding area. Thankfully the effect on humans was very small, with people living close to the reactor receiving an extra radioactive dose equivalent to a standard X-ray.

▶ **Chernobyl, Ukraine, 1986** – caused by a poor reactor design and human error, when reactor operators shut down automatic safety systems. The resulting reactor meltdown caused an explosion and fire which released some of the reactor core material into the atmosphere. Two plant operators were killed by the explosion and 28 people died soon after from radiation poisoning. A 30 km exclusion zone has existed since the accident, and the reactor site is now enclosed in a huge protective steel cover. Ongoing health monitoring of the population surrounding the reactor site has shown no evidence of a major public health impact following the radiation exposure. However, there have been about 1800 confirmed cases of thyroid cancer in children (who were under 14 years of age when the accident occurred). This is a much higher figure than would be expected in the general population, and the psychological effects of the disaster are still an issue for those directly involved.

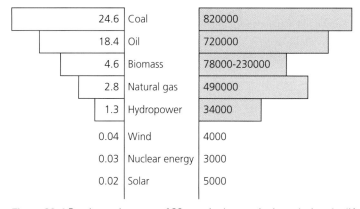

Test yourself

6 What is the main fuel used in a nuclear reactor?

7 What is a 'chain reaction'?

8 Why does a nuclear reactor need a moderator?

9 How can a nuclear reactor in a power station be controlled?

10 Why is the reactor encased inside a steel vessel surrounded by a thick concrete containment structure?

11 Why do spent fuel rods need to be stored under water in ponds within the containment structure?

▶ **Fukoshima, Japan, 2011** – caused by a tsunami wave, which knocked out the coolant pumps, flooded the reactors and led to three core meltdowns with subsequent small explosions. Large amounts of contaminated water were released into the Pacific Ocean. One person subsequently died from radiation poisoning, about 18 were injured in the blasts of the explosions, and about 573 people died as a result of the evacuation. A 20 km exclusion zone was set up and the clean-up is expected to take 30–40 years. There have been minimal long-term health effects on the local population due to the low-level exposure to radiation.

In two of these cases, human error in not following safety protocols was identified as a factor, and in all three, the design of the control and safety systems was at fault. All three disasters have led to major changes in the design and operation of nuclear power plants.

The three major nuclear accidents have impacted on the environment and human health, but this needs to be put into context. Figure 23.4 shows the relative death rates from accidents and air pollution due to the major sources of power (left-hand side). The right-hand side of the graphic shows the relative amount of greenhouse gas emissions from each source. It is fairly obvious that the consequences of using fossil fuels as a power source far outweigh the consequences of using nuclear power.

24.6	Coal	820000
18.4	Oil	720000
4.6	Biomass	78000-230000
2.8	Natural gas	490000
1.3	Hydropower	34000
0.04	Wind	4000
0.03	Nuclear energy	3000
0.02	Solar	5000

Figure 23.4 Deaths and tonnes of CO_2-equivalent emissions during the lifecycle of a power plant per terrawatt-hour of electricity produced.

Test yourself

12 After the Chernobyl disaster in April 1986, a plume of radioactive particles was sent up into the air. The wind carried some of these particles over the uplands of North Wales, where it entered the soil due to heavy rain. Particles of caesium-137 (half-life 30 years) were absorbed by the grass and eaten by sheep. A sample of contaminated soil from a North Wales sheep farm was found to have an activity of 160 Bq in May 1986. Estimate the activity of the sample in May 2046, 60 years after the Chernobyl disaster.

Controlling processes

▶ Nuclear fusion

The energy produced by our Sun (and other stars) is due to nuclear reactions.

In this case, the nuclear reaction involves the **fusion** (joining together) of nuclei, rather than fission. This process produces enormous amounts of energy. 1 kg of burning coal produces 35 MJ (35×10^6 J) of energy. One kilogram of uranium-235 produces 83 TJ (83×10^{12} J) of energy. One kilogram of hydrogen could produce 0.6 petajoules, 0.6 PJ (0.6×10^{15} J) – over 7 times as much as 1 kg of uranium-235!

Even though the Sun produces so much nuclear energy, and fuses hydrogen at a rate of over 6×10^{11} kg/s, the Sun still has enough hydrogen to keep on shining for at least another 5 thousand million years. So, if hydrogen fusion can produce such large amounts of energy, can we develop a nuclear fusion reactor here on Earth to give endless amounts of clean, carbon-free energy? In order to get hydrogen nuclei (protons) close enough to undergo nuclear fusion (and overcome the large force of repulsion due to their positive charge), they need to be moving at very high speed. Because hydrogen is a gas, that means very, very high temperatures (over 15 million °C) and pressures. On Earth these temperatures are very difficult to achieve, and even harder to control and keep going.

Inside the core of the Sun, isotopes of hydrogen (hydrogen-1, and hydrogen-2 (deuterium)), fuse together, making helium-3 and a gamma ray. Two helium-3 nuclei then fuse together making helium-4 and producing two further hydrogen-1 nuclei (or protons). The energy produced by this reaction mainly involves the energy of the gamma rays. A summary of these nuclear fusion reactions is:

$$^1_1H + {}^2_1H \rightarrow {}^3_2He + \gamma$$

$$^3_2He + {}^3_2He \rightarrow {}^4_2He + 2{}^1_1H$$

Key term

Nuclear fusion The joining together of light nuclei at very high temperatures and pressures, releasing large quantities of energy.

✓ Test yourself

13 What is 'nuclear fusion'?

14 Inside the core of the Sun, what are the particles involved in nuclear fusion?

15 Why are high temperatures and pressures needed for nuclear fusion?

16 What is deuterium? How is it different to 'normal' hydrogen-1?

Chapter summary

- The absorption of slow neutrons can induce fission of uranium-235 nuclei, releasing energy, and the emission of neutrons from such fission can lead to a sustainable chain reaction.
- $^A_Z X$ notation equations can be used to summarise nuclear reactions.
- Radioactive materials have an activity (the number of nuclear decays per second), and a half-life (the time taken for the activity of a radioactive sample to halve).
- Moderator material in a nuclear reactor acts to slow down the fast-moving neutrons produced by the nuclear fission process, so that they can cause further fission.
- Control rods are neutron-absorbing rods that can be moved up and down to control the number of thermal neutrons inside the fuel rods.

- Most of the decay products of nuclear fission are radioactive, many of them with very long half-lives, so they have to be carefully stored within the containment structure of the nuclear reactor.
- High energy collisions between light nuclei, especially isotopes of hydrogen, can result in fusion which releases enormous amounts of energy. This is the energy production process in stars.
- Three major nuclear accidents have occurred, and these have involved a failure to follow safety protocols and a failure of control mechanisms.
- Nuclear accidents have consequences for the environment and human health, but these risks need to be compared to the risks associated with power generation from other sources of energy such as fossil fuels.

▶ Practice exam questions

1 In 2021 Cardiff University published findings that using hydrogen peroxide to disinfect drinking water was more effective than using chlorine. Hydrogen peroxide can be easily made and decomposes to form water and another gas, oxygen, O_2.

a) Complete the symbol equation for this reaction. [2]

$$2H_2O_2 \rightarrow \ldots\ldots\ldots H_2O + \ldots\ldots\ldots$$

b) The rate of decomposition is slow at room temperature but can be increased by the addition of a suitable catalyst. Define the term catalyst. [2]

c) Describe a method to investigate the effect of three different catalysts on rate of decomposition of hydrogen peroxide. [6]

2 Slaked lime is made from the exothermic reaction between calcium oxide and water. This chemical is used to extract sugar from sugar cane and sugar beet.

a) Define an exothermic reaction. [1]

b) Suggest how the sugar industry can reduce the risk of a runaway exothermic reaction. [3]

c) Describe the effect of increasing the surface area of calcium oxide. [3]

3 A diagram of a gas-cooled nuclear reactor is shown below.

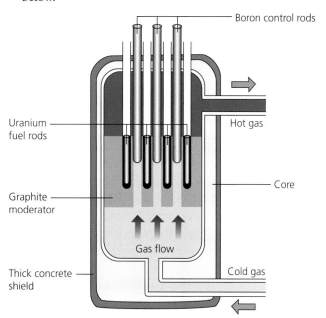

Use information from the diagram to explain how a controlled chain reaction is produced inside the nuclear reactor. [6]

4 Look at the following statements about nuclear fission and fusion. Copy and complete the table by deciding if the statements apply or do not apply (Yes/No) to each type of nuclear process.

Statements	Nuclear fission	Nuclear fusion
Energy is released during the breakup of large nuclei		
A chemical reaction occurs releasing thermal energy		
Energy is released when two light nuclei are forced together		
This is the process that produces energy in the Sun		
This process produces the most energy per kilogram of fuel		
This process can produce a nuclear chain reaction		

Task-based assessment

▶ **Introduction**

This activity-based assessment focuses on three skill areas:

- ▶ devising and carrying out scientific investigations
- ▶ analysing scientific data
- ▶ managing Health and Safety.

The activities that you do in this assessment test the skills that are needed in a laboratory, industrial or commercial workplace. They are set in the context of all the topics that you have studied, **with the exception of:**

- ▶ 2.1.2 Diagnosis and treatment (Single Award)
- ▶ 3.2.1 Processing food (Double Award)
- ▶ 3.4.2 Controlling nuclear reactions (Double and Single Award)

In Double Award you will carry out Activities 1, 2 and 3.
In Single Award you will carry out Activity 1 (plus a risk assessment) and Activity 2.

▶ **Activity 1 – Carrying out a practical investigation in an applied scientific context**

This activity, consisting of three tasks, will be carried out in 3 × 1 hour sessions.

Session 1 – Task A: Planning

In this session, you are required to devise a method to solve a practical problem. The problem will be given to you.

For this task you will be given a list of standard equipment. You need to produce a plan of what you are going to do. Your plan should include:

- ▶ a statement of the variables involved in the experiment (independent, dependent, and controlled)
- ▶ a labelled diagram of your experiment
- ▶ a step-by-step written plan
- ▶ **Single Award only:** a risk assessment for your experiment.

The question paper will prompt you to cover all the parts of the plan. You do this task by yourself, under examination conditions.

Key points

- ▶ Read all the information given to you carefully – there are plenty of clues of what you need to include.

- Make sure you know the different types of variables:
  - independent variable (the one you are changing)
  - dependent variable (the one you are measuring)
  - control variables (the ones you are keeping the same).
- Re-read your method and make sure you have spelt the key scientific words correctly, and used full stops and capital letters accurately.

Session 2 – Task B: Collect and record data

In this session you will carry out your experiment, and measure and record the data as you stated in your plan. You will also be asked to give the resolution of the piece of measuring equipment that you are using (for example, a ruler).

You are given a space to record your rough measurements, plus a space to draw a final table of your results, including a column for any mean average values.

Key points

- Measure and record ALL the measurements needed.
- Make sure you have labelled your table with the correct headers and units.
- Ensure you have used a common number of decimal places in your data on the table.

Session 3 – Task C: Analysis; and Task D: Evaluation

The Task C analysis comes in several formats, but primarily involves plotting a graph or a chart, and then looking for patterns. You are likely to be asked to perform some calculations, usually on extra data that is given to you.

Task D involves evaluating what you have done. You may be asked to:

- assess the suitability of your method
- assess the repeatability of your raw data
- identify sources of inaccuracy
- identify improvements to your method.

Finally, you may also be asked to assess a comment or suggestion made by someone else about the experiment. You could be asked if you agree or disagree with the suggestion, and then to explain your answer.

Key points

- Calculate any mean values correctly and record them with the suitable number of decimal places.
- Plot your graph points accurately on a scale that covers most of the graph paper given.
- When asked to, make sure you explain your answers and give reasons for your choices.

► Activity 2 – Analysing and evaluating secondary data

Activity 2 is a 1-hour session completed under examination conditions. It involves analysing and evaluating sets of secondary data, given to you in a separate Resource Folder. This is very likely to contain a method for doing an experiment, together with some results, usually in numerical form in a table. There is also likely to be some other information, usually presented graphically, either as a diagram, picture, graph or chart.

In Task A, you need to analyse the data supplied in the table. You are likely to be asked to:

- ▸ identify any anomalous data
- ▸ calculate means
- ▸ identify patterns in the data
- ▸ perform some calculations.

You will also be asked to interpret the information presented to you graphically, either by reading values off a graph or chart, or by extracting information from a graphic.

In Task B, you will evaluate the method outlined in the Resource Folder. You may be asked to:

- ▸ assess the suitability of the method
- ▸ comment on repeatability
- ▸ comment on potential sources of uncertainty.

Finally, you may also be asked to assess a comment or suggestion made by someone else about the experiment. You could be asked if you agree or disagree with the suggestion, and then to explain your answer.

Key points

- ▸ Remember that anomalous data does not fit the general pattern of the rest of the data and should be excluded from any analysis.
- ▸ Always use the same number of decimal places as the rest of the data for calculated means.
- ▸ You will need to read the given method very carefully in order to spot potential sources of uncertainty.

► Activity 3 – Managing Health and Safety

This is a 1-hour session, completed under examination conditions, where you are given details of an experiment carried out in an Applied Science context. The experiment is likely to feature several potentially hazardous methods; pieces of apparatus; or chemicals; and you will be given the relevant CLEAPSS Student Safety Sheets for the experiment. Your task is to produce a risk assessment for each part of the experiment. **You will not carry out this experiment.**

In the risk assessment you should identify:

▶ **Hazards** (**things** that can **harm** you) AND state the nature of the hazard (why it is **harmful**); for example, a Bunsen burner flame is hot.

▶ **Risk** (the **nature of any injury** that could occur (including **part(s) of the body** that could be injured) AND the **action** that produced it; for example, the hot water could scald my hand while I am pouring it.

▶ **Control measures** (steps that you could take to **minimise the risk**); for example, wear goggles or ensure hair is tied back.

Some obvious hazards/risks/control measures may be given to you, or partially given, and sometimes there may be more than one control measure for a hazard.

Key points

▶ Study the method and the diagram for the experiment carefully to identify all the hazards.

▶ Always read and check the Student Safety Sheets, particularly for hazardous chemicals, as the sheet will give you the nature of the hazards and why they are harmful.

▶ The control measures that you pick must be relevant to the hazard and risk identified.

Practical assessment

▶ Introduction

In this assessment you will be asked to show your ability to work in a scientific way. You will need to use your practical skills to obtain data from a given experimental method, and then analyse and evaluate it.

▶ Double award students need to complete **TWO** tasks.

▶ Single award students only complete **ONE** task.

Each task is split into two sections, the experimental phase and the analysis of results, with each section lasting 60 minutes.

▶ Section A – Obtaining results

In this section of the task(s), you will be given an experimental method and you will work in a group of two or three, to obtain results. An example of an experiment could be:

'Investigate the rate of cooling of an uninsulated conical flask.'

Although you carry out the experimental work as a group, you must answer the questions on the examination paper by yourself. You will be asked to:

- complete a risk assessment
- write a hypothesis for the experiment
- construct and complete your own table of results.

Key points

- Sometimes apparatus does not work, or it breaks. If this happens, make sure that you ask your teacher for help. They can give you replacement equipment, materials or chemicals.
- Remember to identify the nature of any hazards and the actions associated with the risks.
- Make sure that each column of your results table has a header and the correct units.
- Do not put units in the body of a results table.

▶ Section B – Analysing and evaluating results

This section of the assessment is carried out under examination conditions. You are normally asked to:

- identify the independent, dependent and controlled variables in the experiment you carried out
- plot a graph of your data (or very similar data)
- identify the pattern in the data (or identify no pattern)
- compare the pattern in your data to your hypothesis
- calculate a value from your results
- evaluate the quality of your data, including:
  - **uncertainties** – the intervals within which the true value is expected to lie.
  - **accuracy** – a measurement result is considered accurate if it is judged to be close to the true value.
  - **precision** – the closeness of agreement between measured values.
  - **improvements** – steps that you can do to reduce uncertainties.
  - **repeatability** – a measurement is repeatable if repetition by the same student or group of students using the same method and equipment obtains the same or similar results.
  - **reproducibility** – a measurement is reproducible if repetition by different students or groups of students obtains the same or similar results.

Key points

- Remember – the **independent** variable is the one you change; the **dependent** variable is the one you measure; and the **controlled variables** are those you keep constant.
- Plot graphs so that the scale fills most of the area of the graph paper, generally with an origin that starts at (0,0).
- Best fit lines can be curves or straight lines and should be drawn as smooth lines.
- Make sure that you know the definitions of the terms: uncertainty, accuracy, precision, repeatability and reproducibility.

Glossary

Absorption The movement of food molecules from the gut into the bloodstream.

Absorption spectrum The pattern of black lines in the spectrum of light from a star that shows the presence of different elements in the atmosphere of the star.

Acid A soluble substance that releases $H^+(aq)$ in solution.

Activation energy The minimum amount of energy needed to start a chemical reaction.

Active site The place on an enzyme molecule where the substrate attaches.

Aerodynamics Minimising the air resistance by changing the shape of the vehicle/object, so that air flows smoothly over the surfaces.

Alkali A soluble base that releases $OH^-(aq)$ in solution.

Allotropes Different physical forms of the same element.

Amino acid Chemical group from which proteins are formed.

Anomalous result A result that does not fit in the pattern of the other results.

Atmosphere The envelope of gas surrounding our planet.

Atom The smallest particle that can exist on its own.

Atomic number Number of protons in the nucleus of an atom.

Bacterial spores Highly resistant, dormant forms of bacteria, which are formed in response to adverse environmental conditions.

Base A substance that reacts with an acid.

Biodegradable Capable of being broken down by microorganisms in the environment.

Biodiversity The variety of living organisms in an area.

Biological control The use of natural predators to control pests.

Biomass Dry organic matter made from dead organisms.

Bioplastic Biodegradable plastic made from biomass feedstocks.

Blast furnace Tower where iron ore is reduced using carbon. Hot air is blown in.

Brittle Materials do not stretch before they fracture.

Bulk properties Properties of a large (hand-held) piece of the material.

Carbon footprint The equivalent mass of carbon dioxide gas produced when an energy source generates electricity.

Carrier Individual with a recessive allele for a gene. The characteristic determined by the allele is not shown, but can be passed on to the next generation.

Catalyst Chemical that increases the rate of a reaction by reducing the activation energy without itself undergoing any permanent chemical change.

Categoric data Data that is a category, usually a word, e.g. eye colour.

Chain reaction When one nuclear fission produces several neutrons which go on to produce further fissions, which also go on to produce further fissions, and so on.

Chemical reaction A change where a new substance is made and mass is conserved.

Chromosome Thread-like structure made of DNA, found in the nucleus of cells.

Clinical trial Involves testing a new drug on volunteers.

Colorimetry A quantitative technique that can be used to determine the concentration of a solution based on its transmission of light through the sample.

Competition A relationship between organisms (of the same or different species) where they require a resource that is in limited supply (for example, food, light or water).

Compound Substance made of two or more different types of atom chemically joined.

Concentration gradient The difference between two concentrations.

Conduction The transfer of heat energy from hot to cold by the vibration of particles within solids and liquids.

Continuous data Data that can take any value, e.g. height, handspan, temperature.

Control rod A nuclear reactor rod of material, such as boron, that absorbs neutrons. Lowering or raising the control rods inside a reactor can control the rate of the nuclear fission reactions.

Convection The transfer of heat energy from hot to cold by the movement of particles through liquids and gases.

Correlation A connection between two or more things, so that when one of them changes, the other also changes in a predictable way.

Covalent bonds Bonds formed between atoms that share electrons.

Cracking Decomposition of long chain hydrocarbons to make shorter more useful hydrocarbons for fuels and reactive small hydrocarbons to make polymers.

Crop rotation The practice of growing a series of different crops in the same area in successive growing seasons, avoiding exhaustion of soil nutrients; peas and beans actually add nitrates to the soil.

Data Information, for example from observations or measurements.

De-oxygenated blood Blood containing a low level of oxygen.

Digestion The breakdown of food molecules into small, soluble molecules.

Discontinuous data Numerical data that can only have specific values and no intermediate values, e.g. shoe size.

DNA Deoxyribonucleic acid – the chemical from which genes are made and which controls the production of proteins in cells.

DNA profiling An analytical technique used to determine an individual's characteristics.

Double circulatory system Blood system in which the blood travels through the heart twice on each circuit of the body.

Drug A chemical that alters the way that the body works in a certain way.

Durable Can withstand wear, pressure or damage.

Ecosystem A community or group of living organisms together with the habitat in which they live, and the interactions of the living and non-living components of the area.

Effervescence Seeing bubbles and/or hearing fizzing.

Efficiency The ratio of energy (or power) usefully transferred/total energy (or power) supplied, which is normally expressed as a percentage.

Egestion The passage of undigested materials out of the body.

Electrical-characteristic graph Current–voltage graph.

Electrolysis Electricity is used to decompose an ionic substance into simpler substances.

Electrolyte A solution containing ions.

Electromagnetic spectrum A family of waves that all travel at the same speed, c, the speed of light (3×10^8 m/s in the vacuum of space).

Electrostatic attraction Ionic bonds are formed by this attraction between oppositely charged particles (ions).

Element Substance that cannot be broken down by chemical means.

Emulsification The breaking up of large droplets of liquid into smaller ones.

Emulsion A mixture of two or more liquids, in which one is present as microscopic droplets and is distributed throughout the other.

Endothermic Process takes in energy.

End point The point at which the indicator has changed colour in an acid–base titration.

Energy efficiency Rating of appliances is on an A–G comparison scale, where A-rated devices are more efficient in their use of energy than G-rated devices.

Enzyme Biological molecule which acts as a catalyst, speeding up a chemical reaction but not taking part in it.

Enzyme–substrate complex An enzyme and its substrate(s) joined together.

Exothermic When energy is given from the system to the surroundings.

Filtrate The liquid collected in the conical flask when a mixture has been filtered.

Finite resource A non-renewable resource that is being used faster than it can be created.

Fossil fuels A finite energy source made from ancient biomass.

Fractional distillation Physical separation technique used to separate components of a solution based on their different boiling points.

Galaxy A distant collection of stars in space, orbiting around a common centre of gravity (usually a massive black hole).

Gamete Sex cell (egg or sperm in animals, pollen and ovule in plants).

Global warming The gradual increase in overall average global atmospheric temperature.

Gradient The size of the slope of a straight line on a graph.

Group A column of elements in the Periodic Table.

Habitat The place where an organism lives.

Haemoglobin Red pigment in red blood cells that carries oxygen.

Hard water Water with dissolved calcium or magnesium salts that does not foam with soap.

Heritable Capable of being inherited (because it is a result of genes).

Homeostasis The maintenance of a constant internal environment.

Inbreeding The breeding of individuals which are closely related and so share many similar alleles.

Indicator species A species with a known tolerance (high or low) to a particular pollutant, which can be used to indicate the level of pollution in an environment.

Infrared radiation Electromagnetic radiation that we can feel as heat.

Insulation Home insulation systems reduce the heat energy losses from a house.

Ion A charged atom or group of atoms.

Ionic compounds Formed between particles joined by ionic bonds.

Isotope Atoms of the same element with the same atomic number but different mass number.

Kilowatt hour The unit of electrical energy consumed.

Mass number Number of protons and neutrons in the nucleus of an atom.

Medical imaging Waves are used to image inside the body to reveal, diagnose or examine injury or disease, without the need for surgery.

Metabolism All the chemical reactions going on inside a living organism.

Metal An element on the left or centre of the Periodic Table or an alloy.

Metallic bonds Form when metallic positive ion cores get arranged in a lattice structure, through which a 'sea' of delocalised electrons can flow.

Miscible A liquid that can mix with water and form a solution.

Moderator Material such as water that slows down neutrons in a nuclear reactor so that they can produce further fissions.

Mole The amount of a substance; one mole of any substance contains 6.02×10^{23} particles.

mol dm-3 Moles per cubic decimetre is the unit of concentration. 1 $mol\ dm^{-3}$ means that there is 1 mole of a substance dissolved in 1 dm^3 (1 litre) of solvent.

Monomer Small reactive molecules, usually with C=C, that can join together.

Multicellular Consisting of more than one cell.

Neutralisation A chemical reaction between an acid and a base to make a salt.

Nitrogen fixation The conversion of nitrogen in the air into nitrates.

Nuclear fission When a large unstable nucleus breaks up spontaneously, or on impact with a neutron, releasing energy.

Nuclear fusion The joining together of light nuclei at very high temperatures and pressures, releasing large quantities of energy.

Nucleus The massive positively charged centre of an atom.

Oxidation Chemical change where oxygen is added or electrons are removed.

Oxygenated blood Blood containing a high level of oxygen.

Pathogen An organism that causes disease, such as a virus, bacteria, fungi, or parasite.

Period A row of elements in the Periodic Table.

Photosynthesis The process by which plants make glucose using carbon dioxide, water and light energy.

pH scale A measure of acidity of a solution on a scale from 0 to 14.

Placebo A version that looks exactly the same as the real drug, but has no effect on the human body.

Planet A spherical object in space, orbiting a star, which has enough gravitational strength to clear other objects out of its orbit.

Polymer Long chain, organic, covalently bonded molecule made up of many repeating 'mer' molecular units.

Power The rate at which a device transfers energy from one form into other forms.

Precipitate An insoluble solid produced during a chemical reaction in solution.

Punnett square Table which shows the possible crosses of gametes in a genetic cross.

Pure A substance containing only one type of particle.

Radiation The transfer of heat energy from hot to cold by the transmission of infrared electromagnetic waves.

Rate of a chemical reaction The speed of a chemical change.

Raw material Unprocessed material from land, ocean or air.

Reaction pathway A series of different steps that happen in a chemical reaction to get from the reactants to the products.

Reduction Chemical change where oxygen is removed or electrons are gained.

Relative atomic mass The weighted average mass of an atom considering the isotopes available.

Relative molecular mass The sum of the relative atomic masses of all the atoms in a molecule.

Renewable Energy sources produced by the action of the Sun and not used up when working.

Residue The solid collected in the filter paper when a mixture has been filtered.

Retention factor The ratio of how far a substance has travelled compared to a solvent in the same medium.

Rolling resistance The effect of friction between a vehicle's tyres and the road surface, which reduces the efficiency of the vehicle.

Salt A neutral ionic compound produced from a neutralisation reaction.

Salting Adding salt to food, either on the surface or in water surrounding the food.

Sankey diagram A diagrammatic way of showing energy transfers. The wider the bar of the diagram at any point, the bigger the energy being transferred.

Selectively permeable membrane A membrane which allows some substances to pass through but not others, sometimes called a semi-permeable or partially permeable membrane.

Side effect The negative effect of a drug on the body.

Star An object in space emitting electromagnetic radiation due to the nuclear fusion of hydrogen and other light elements.

Stoma (pl. stomata) Pore in a leaf which lets carbon dioxide into the leaf and lets water vapour out.

Supernova The huge explosion that happens when a massive giant star runs out of nuclear fusion material and implodes (collapses in on itself).

Sustainability Using renewable energy and using that energy very efficiently.

Symptom A condition of a disease, caused by the disease, e.g., a fever is a symptom of flu.

Tangent The straight line that best represents a small section of a curved trend line on a graph.

Tectonic plates Seven or eight very large slabs of rock that make up the Earth's crust and float on the mantle.

Thermal decomposition The breaking down of a substance into simpler substances using heat.

Thermal neutrons Neutrons produced inside a nuclear reactor that are slowed down by a moderator so that they can produce further nuclear fissions.

Thermal runaway Can occur in an exothermic reaction, when increasing the temperature increases the rate of reaction, which then increases the temperature further.

Titre The total volume of substance added from the burette (titre = final volume reading on the burette – initial volume reading on the burette).

Toxin Poisonous substance.

Turbidity A liquid that is cloudy or opaque.

Ultrasound Sound waves with frequencies higher than the upper limit of the human hearing range.

Index